Applications of Microorganisms in Industrial Biotechnology

Applications of Microorganisms in Industrial Biotechnology

Patrícia Alexandra Batista Branco

Delve
PUBLISHING

www.delvepublishing.com

Applications of Microorganisms in Industrial Biotechnology

Patrícia Alexandra Batista Branco

Delve Publishing
2010 Winston Park Drive,
2nd Floor
Oakville, ON L6H 5R7
Canada
www.delvepublishing.com
Tel: 001-289-291-7705
001-905-616-2116
Fax: 001-289-291-7601
Email: orders@arclereducation.com

ISBN: 978-1-77361-201-0 (Hardcover)

ABOUT THE AUTHOR

Patricia Alexandra Batista Branco has received her PhD in Food Engineering from Higher Institute of Agronomy, Lisbon, Portugal and Masters' in Molecular Biology in Health from Egas Moniz School of Health, Portugal. She is an active researcher with many years of experience and has published research papers with reputed Journals.

TABLE OF CONTENTS

PREFACE

The field of microbiology and biotechnology are intertwined since time immemorial however, the ties between the two areas became prominent in the last century. The areas provided various products which enriched mankind in various ways mainly in the form of food and succeeded in producing medicines. There was no technology which provoked the humans to understand the mechanisms involved whilst using microorganisms. In previous millennia, microorganisms were utilized by humans for several needs; however, there was no scope of understanding the machinery to the complete detail.

The nineteenth century bore an outstanding scientist named Louis Pasteur who pioneered in industrial microbiology. His understanding of microorganism laid a path to the other discoveries which made human life more comfortable and also increment in life span is clearly noticed. The fight against infectious diseases has progressed with the advancements in microbiology. The era of mass production of the microbial products initiated mainly with citric acid production. The Second World War provided an essentiality to understand the process of preservation of products in aseptic conditions.

The economically viable products such as vaccines, cytokines, pharmaceuticals and foods were produced in a large-scale due to advancements in genetic engineering in the seventies. The applied microbiology and biotechnology are playing a crucial role in dictating national economy, medicine, agriculture, environmental protection and pharmaceuticals. The main reason to devise this part of literature is to introduce and summarize the current state of knowledge which concerns microbial application in large scale production lines. This book is built on my experiences with several research fronts during these two decades.

The field of industrial microbiology and biotechnology deals with exploitation of microorganisms are a systematic manner in order to obtain goods and services for human welfare. The two-immediate aspect of industrial microbiology are fermentation processes and service delivery especially in pollution control. It is assumed that the reader may have got some learned experience in microbiology to understand this book. The students of any life sciences and chemistry can understand the concept delivered in this book without any hassles.

The application of microbiology in industrial biotechnology is broadly emphasized in this book. The chapters were designed to let the reader take a systematic study without getting struck at any concept and never feel confused.

I would like to express my gratitude to all the professors and researchers who provided me variety of inputs to make this literature work a success. All the valuable time they invested in me to bring out this book is duly appreciated and some of the reflections which they expect are in due till the book is read by many of the enthusiastic students.

Author

CHAPTER 1

THE DIVERSITY OF MICROORGANISMS

INTRODUCTION

In ancient Egypt, Fermented bread was produced six centuries ago and it was evidenced by the artifacts found in the pyramids. Dionysos is one of the Greek gods who was believed to teach the process of fermentation of fruits. Brewing of beer was documented three centuries ago in Babylonia. In Central Asia and the Balkans milk beverages were fermented and sold fifteen hundred years ago so as the China and Japan in fermentation of soybean. About 3000 years ago, the curd produced from soya bean fermentation was used to cure skin infections in China. In Central America, Indians used fungi to treat wounds. In Sudan, about 1400 years ago, tetracycline was used by Nubian farmers who harbored Banks of Nile. It is believed that the farmers produced tetracycline from Streptomyces species which infected cereals. Libavius was considered as the first scholar to work on fermentation and he also explained the variation between putrefaction and fermentation. In the year 1669, a scholar named Becher explained the main difference in the processes of putrefaction and fermentation and he also explained that the materials which are rich in saccharides have capability to produce wine spirit. Erxleben in the year 1818 mentioned that the microorganisms observed by Antonie Van Leeuwenhoek were responsible for fermentation process. Gay Lussac in the year 1815 formulated the reactions behind the fermentation of alcohol. Louis Pasteur in the year 1857 showed that the fermentation of alcohol is due to yeast. The process of sterilization during microbial culture isolation in pure form was devised by Robert Koch. Citric acid, propanol, butanol, acetone and glycerol were produced using microorganisms during First World War. In Second World War, antibiotics were discovered. Most important revolution was the process of cell immobilization and enzyme immobilization which had drastically reduced the biotransformation costs.

Industrial microbiology can be defined as the application of principles in science and engineering in transforming the raw materials via microbiological process in order to produce goods and provide services. Moser in the year 1992 explained goods as the food and pharmaceutical products whereas services refer to the treatment of water and also processing of agricultural and industrial wastes. The major fields which route the field of industrial microbiology are biochemistry, physiology, microbial genetics and general microbiology. Some of the disciplines linked with science and engineering are Microbiology for Taxonomy and ecology, Biochemistry for biocatalysts such as enzymes and cells, Genetics for improvement and selection, Physiology for growth proliferation and product formation, Molecular biology for regulatory mechanisms, Microbial engineering for modeling via mathematical processing, Mechanical engineering for isolation and culture devices, Chemical engineering for Mass transfer and oxygen transfer, and Price engineering for studying economics of raw materials and energy.

The species used for production at industrial scale are still less in number. This is due to excess use of traditional microorganisms, use of traditional models while doing research in the fields of biochemistry and genetics, existence of negligence of some key microorganisms such as slow growing and anaerobic bacteria and fungi. One of the key discoveries which revolutionized the field of industrial microbiology is the microbial production of active compounds which were used in case of gastric ulcers, tumors, overweight, thromboses and hypertension.

Enhanced knowledge in primary and secondary metabolisms gave an impetus to study the effective utilization of bioreactors. The enzymes were costlier and it pushed the barriers of technology.. Enhanced metabolite production was achieved by using industrial microorganisms via random mutations and selection instead of traditional knowledge of biochemistry and genetics. The technique of cell fusion and protoplast fusion along with DNA recombination allowed the cross among the species and improvement of yield. Protoplast fusion was choice of concern where large genomes are involved. The limited interest into the aspects of engineering in microbial processes was due to the exaggerated advances in other areas such as genetic engineering rather than microbiological engineering.

The cell's nature is of top most important characteristic which is to be observed by a microbial biotechnologist. The cell wall is present in all prokaryotic cells whereas is only present in the eukaryotic cells of fungi and plants. .. The cell membrane is present in both prokaryotes and eucaryotes and is basically a double layer enriched with phospholipids and it is the envelope where the cell is enclosed. It helps the cell in absorption and also excretion. The proteins are synthesized from ribosome. The ribosome is composed of 2 units which are often called as sub-units. The ribosomes present in prokaryotic and eukaryotic cells are designated as 70S and 80S, respectively.

The smaller sub-unit is called 30S and 40S sub-unit and the larger sub-unit is called as 50S and 60S sub-unit for prokaryotic and eukaryotic cells, respectively. S refers to Svedberg units. The Svedberg unit is nothing but the rate of sedimentation observed in an ultracentrifuge. The rate of sedimentation is directly proportional to the particle size. The Svedberg units should never be added. For prokaryotes, the 30S sub-unit is produced from 16S RNA molecule with the help of 21 chains of polypeptides. The 50S sub-unit is produced from two types of RNA molecules. They are 5S and 23S RNA molecules and 34 chains of polypeptides. The mitochondria is only present in eukaryotic cells and is considered as the power houses of the cell. Mitochondria are membrane bound structures. The eukaryotic cell living in aerobic conditions utilizes mitochondria in respiration process mainly in oxidative phosphorylation. The prokaryotic cells do not contain mitochondria and the respiration generally takes place via cell membrane. The nucleus is enclosed by nuclear membrane in eukaryotic cells. The nuclear membrane is absent in prokaryotic cells. A single DNA molecule which is circular is generally present in prokaryotes and it forms the genome or the hereditary apparatus of the cell. In eukaryotes, the whole DNA is distributed into chromosomes. The eukaryotic nucleus contains a specific structure called as nucleolus. The nucleolus is responsible for the production of ribosomal RNA. The ribosomal proteins are generally produced in the cytoplasm and they get migrated into the nucleolus for and unite with ribosomal RNA (rRNA). In the nucleolus, small and large units of ribosomes are assembled in eukaryotes. After proper assembly, they get exported into the cytoplasm.

Whittaker's system of classification of living organisms is well accepted all over the world. The classification is primarily based on cell type i.e. prokaryotic and eukaryotic, organizational level i.e. multicellular or single cell and the type of nutrition i.e. autotrophy and heterotrophy. Based on these key characteristics Whittaker classified living organisms into five main groups. They are Monera which contains bacteria, Protista which contains algae and protozoa, Fungi, Plants and Animals. A present the classification followed by many is proposed by Carl Woese. The classification is prepared as per the rRNA sequences present in 16S sub-unit often referred to as small sub-unit or SSU in case of ribosomes in prokaryotes. In eukaryotes, the SSU called 18S is also widely studied and it formed basis for present classification.

The main advantages while using rRNA sequence are that 16S and 18S RNA is universally distributed and they are important for the formation of ribosomes, the function of the sub units is same in almost all ribosomes and the sequence is well conserved and it is useful to compare distance and closely related species.. As per the newly framed classification system, the living organisms are classified into Archae, Bacteria and Eukarya.

Microorganisms have some beneficial qualities which make them the best choice over animals and plants. The time taken for an organism to get maturity and able to reproduce varies and it is called as regeneration time. The regeneration time in humans is 12 years, in cattle it is two years, one and half year in pigs, six months in chicken and it takes just 15 minutes for *E. coli* which is a bacterium. The products obtained will take just days with microorganisms but with animals and plants, it takes months to years. The microorganisms are not affected by weather but plants and animals get affected. The microorganisms are almost resistant to diseases but there are some contaminants and phages which attack them.

As per the Bergey's manual, Bacteria was splitted into 18 phyla. Proteobacteria was considered as the most advanced phylum. The most evolutionary phylum is Aquiflex. The phyla which are used extensively in biotechnology and industrial microbiology are Actinobacteria, Firmicutes and the Proteobacteria. The Proteobacteria constitutes major members of bacteria. The diversity was common among these members so they were named after Greek god named proteus who could change his shape. This phylum constitutes nitrogen fixing bacteria, *Heliobacter*, *Vibrio*, *Salmonella* and *Escherichia*.

All the members of this phylum are generally Gram negative with lipopolysaccharides in the outer membrane. Proteobacteria was classified into Alpha (α), Beta (β), Gamma (Y), Delta (δ) and Epsilon (ε). The main organisms which are extensively used in industries are *Gluconobacter* and *Acetobacter.* They are called as acetic acid bacteria and are members of Alphaproteobacteria. *Acetobacter* and *Gluconobacter* produce acetic acid via partial oxidation of alcohol and it is widely used in the production of vinegar. *Gluconobacter* do not have the citric acid cycle so do not oxidize acetic acid. Acetobacter oxidizes acetic acid into carbondioxide. They can survive in acidic pH as low as 5. The property of partial oxidation of sugars was exploited for industrial use. The main uses are the glucoronic acid production from glucose, arabonic acid produced from arabinose and galactonic acid produced from galactose. The sorbose production from sorbital by using acetic acid bacteria is an important step in mass production of Vitamin-C or ascorbic acid.

Firmicutes is another bacterial division dominated by gram positive bacteria. Mycoplasmas are the members of this division and they do not have cell walls and hence cannot respond to Gram staining. They possess low concentration of Guanine and Cytosine Nitrogen bases in their DNA. The G+C ratio is used to classify bacteria according to the taxonomical framework. It is nothing but the ratio of GC to Guanine, Cytosine, Adenine and Thymine of the bacterial cell. The GC ratio is G+C upon G+C+A+T and multiplied by 100. The GC ratio is essential in classifying gram-positive bacteria. Fermicutes contain gram positive bacteria with low GC content i.e. less than 50% whereas Actinobacteria shows G+C concentration more than 50%. The industrially important microorganisms of Firmicutes were classified into spore forming, non-spore forming and cell wall-less organisms.

The group of Firmicutes which are spore forming are further classified into *Bacilllus spp* and *Clostridium spp*. The Bacillus spp are mostly aerobic whereas the Clostridium spp are mostly anaerobic in nature. *Bacillus thuringiensis* possess certain genes which are responsible for toxin production and these genes were successfully inserted into plant genome making the plant insect resistance. A lactic acid bacterium is one of the members of non-spore forming Firmicutes. These members contain low G+C content and are considered very important for industrial use. These bacteria are cocci and rod shaped. Some of the genera which belong to this group are *Streptococcus*, *Pediococcus*, *Leuconostoc*, *Lactococcus*, *Lactobacillus* and *Enterococcu*s. Substrate level phosphorylation is dominant in these members because they lack cytochrome and porphyrins and hence unable to conduct electron transport chain. The bacteria of this group have no ability for speedy growth and hence addition of nucleotides, vitamins and amino acids is mandatory.

The members of lactic acid bacteria are classified into homofermentative group and heterofermentative group. The homofermentative group produces only lactic acid as a result of fermentation of sugars. The heterofermentative group produces ethanol, lactic acid, and carbon dioxide. The main characteristic observed in heterofermentative organisms is the absence of aldolase. This enzyme is responsible for the splitting of glucose into three sugar molecules via Embden–Meyerhof–Parnas (EMP) pathway. Glyceraldehyde-3-Phosphate is converted into lactic acid by lactic acid bacteria which is homofermentative. Xylulose-5-phosphate is produced from pentose pathway in heterofermentative lactic acid bacteria. The Xylulose is a five carboned compound and it gets splitted into glyceraldehyde-3-phosphate which is a 3 carboned molecule. This leads to the formation of lactic acid and acetyl phosphate which is two carboned compound. The acetyl phosphate gets converted into ethanol.

The lactic acid bacteria can ferment raw materials rapidly. They use less nitrogenous compounds. They can produce maximal yields of lactic acid. They can grow in high temperature and low pH. They produce minimal byproducts and lesser cell mass. The selection of lactic acid bacterium depends on the carbohydrate which is to be fermented. Sucrose can be fermented by *Lactobacillus delbreuckii*. *Lactobacillus helveticus* can use galactose and lactose whereas *Lactobacillus delbreuckii* uses lactose. *Lactobacillus amylovirus* and *Lactobacillus amylophylus* can ferment starch. Sulfite waste liquor can be fermented by *Lactobacillus pentosus*.

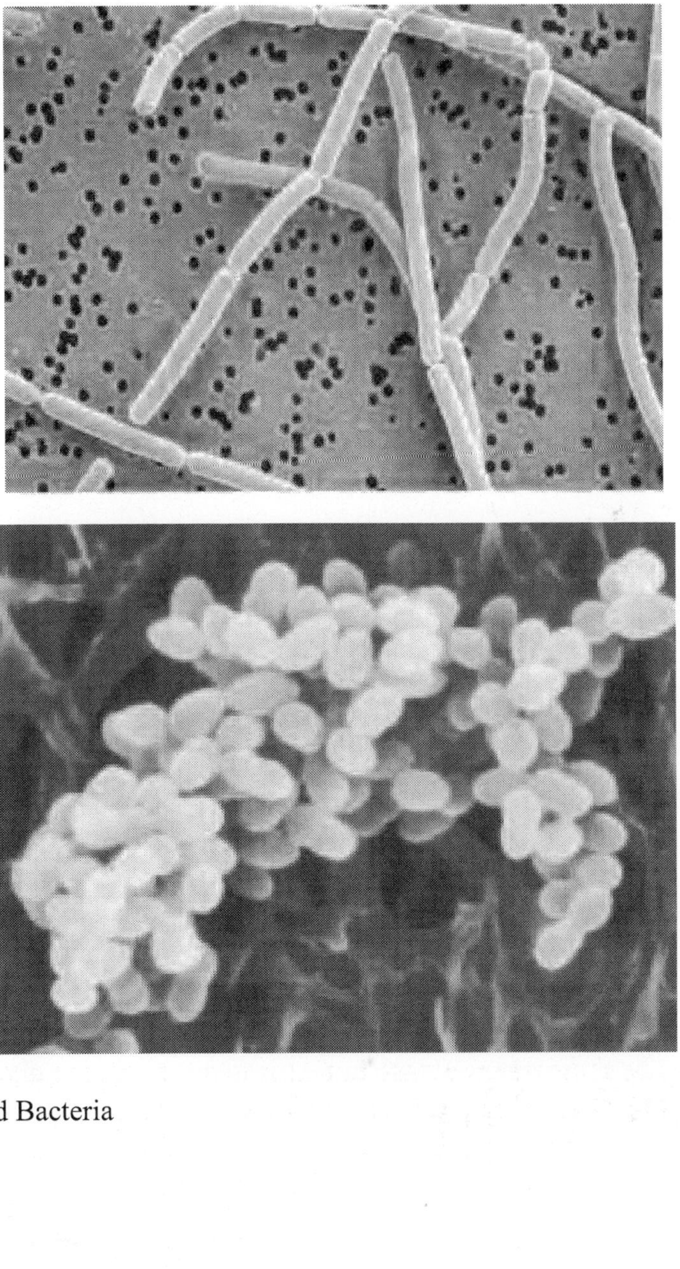

Lactic Acid Bacteria

The Actinobacteria belongs to the group of Firmicutes which has GC content greater than 50%. The name Actinobacteria is due to their capacity to form hyphae or filaments. The members of Actinomycetes bear filamentous hyphae. They contain peptidoglycan molecules in their cell wall and their size is 1μm in diameter. Actinomycetes are useful in production of pharmaceuticals. Streptomyces can produce antibiotics and it has the capacity to produce drugs which are not antibiotic. The members of Actinomycetes dominate soil.

Eukarya includes certain members which are of good use in the field of biotechnology. The fungi which is a member of Eukarya was classified into Basidiomycetes, Fungi imperfecti, Ascomycetes and Phycomycetes. *Mucor* and *Rhizopus* belong to Phycomycetes or Zygomycetes and they are capable of enzyme production. Ascomycetes contain Yeast as a member and they produce alcohol beverages and ethanol. Ergot alkaloids are produced by *Claviceps purperea*. Penicillin is an antibiotic produced from *Penicillium* species. Aflatoxin is a type of food toxin produced from *Aspergillus*. *Aspergillus* and *Penicillium* are the members of Fungi Imperfecti. Mushrooms are the edible fruiting bodies produced by Agaricus. Agaricus belongs to Basidiomycetes.

The microorganisms which are to be used for production on an industrial scale needs to satisfy certain criteria of requirements. Some of the requirements are

- The microorganism should be capable to grow in a minimal medium and it should require a negligible amount of growth factors such as acids, nucleotides and vitamins. One should bear in mind that the excess use of additional supplements will increase the fermentation cost and hence the commodity being produced..
- The growth of the selected microorganisms needs to be rapid with the minimal medium one could provide.
- It should produce the products of interest and they might be metabolic products or cell itself.
- The toxic materials are not expected to be in the end products. This rule needs to be strictly followed when the products are to be consumed for internal use.

- The selected microorganisms needs to be genetically stable so that the physiology can be stable. It would be a costly risk if the microorganisms mutates inadvertently. The mutation in the microorganisms can yield undesired or toxic product.
- The selected microoorganisms need to provide an efficient method to harvest the products at the end of the production process. The bacteria and yeast can both be equally used for production on an industrial scale. The better option in this case would be to use yeast as it will be easy to recover via centrifugation. The diameter of the bacteria is approximately 1μm and the diameter of the yeast is approximately 5μm. Supposing same densities for both the organisms, yeast could sediment 25 times faster than that of bacteria. The rapid sedimentation can result in minimal expenditure in terms of power consumption, limited supervision, and hence high rate of profits.
- The candidate organism should satisfy the physiological requirements needed for its own protection from the contaminants. A microorganism which has the capacity to grow in low pH and in high temperatures is considered as advantageous. Mesophiles are less beneficial over the thermophiles.
- The microbe selected should exhibit resistance towards *Bdellovibrio* spp which is considered as predator. There should not be any bacteriophages. It is beneficial to use phage-resistant and high yielding organisms.
- Aeration in the fermentation constitutes about 20% of the production cost. It is logical to use low oxygen demanding microorganisms.
- Though the organism which is selected is resistant to mutation, but still the microbe needs to be amenable for a manipulation at genetic level for beneficial changes.

MICROBIAL DIVERSITY

The phylogenies classify microorganisms into three domains namely bacteria, Archaea and Eukarya. The Eukarya includes protists, fungi, plants and animals. The virus is positioned in an uncertain position as far as the phylogenetic tree of life is concerned. The scope of this book lies with bacteria, Archaea and Fungal members and their association with biotechnology at an industrial scale. We will also learn the role of viruses in generating devastating problems in industries. The Bacteria and Archaea domains exhibit diversity in several aspects. They possess varied source of energy and also the source for generating cell carbon and nitrogen which is different among the species. Variations do exist in their metabolic pathways, end products and the power to absorb natural compounds. Barophiles are the microorganisms which can live in extreme pressures, mostly in sea depths. Halophiles love to live in salt concentrated areas and thermophiles lives in high temperatures. The bacteria can exist as symbionts and also can live as parasites in mammalian cells. The enormous diversity of microorganisms existent does provide limitless opportunities in industrial biotechnology.

Fungi produce enzymes which have the capacity to break biopolymers such as lignin, polysaccharides and proteins. Numerous tiny organic molecules with not usual structure are being extracted from the fungal members. The molecules of primary importance are antibiotics. Some properties of bacteria are better than fungi. Fungi lack the machinery needed for nitrogen fixation and photosynthesis and they cannot absorb inorganic compounds readily. Consortium do exist among bacteria and fungi so that they can benefit from each other. An estimate suggests that 90% of the carbon dioxide produced is due to the respiration of bacteria and fungi. Human build technology utilizes the outstanding abilities of microbial mixture in daily fermentations, food, beverages and also in specialized processes such as biotreatment of waste water. The degradation of oil spills on a mass scale by microorganisms attracted attention from many environment lovers. The key advantage of biodegradation is that negligible amount of damage is seen in the local environment. This chapter will provide certain information about microorganisms and the necessity to explore the microbial diversity for biotechnological applications.

Classes

The organisms are classified into two main classes called prokaryotes and eukaryotes based on the variations observed in the interior anatomy of their respective cells. The eukaryotes have a nucleus and it is membrane bound. The nucleus contains all the genetic information required for the cell being in state of growth. The eukaryotes mitochondria which is a membrane bound organelle and carry genetic information. The prokaryotes contain crude nuleoid which is devoid of membrane and mostly circular DNA is observed. The bacteria and Archaea members are considered as prokaryotic and fungi are eukaryotic. The comparison between prokaryotes and eukaryotes is illustrated in table-1. *Saccharomyces cerevisiae* is one of the fungus of choice for a particular application so as the bacterium named *Escherchia coli.*

Table 1: Comparison chart of Bacteria, Archaea and Eukaryotes

	Bacteria	**Archaea**	**Eukaryotes**
Structural features			
Chromosome number	01	01	>1
Nuclear membrane	None	None	Present
Nucleolus	None	None	Present
Mitotic apparatus	None	None	Present
Microtubules	None	None	Present
Membrane Lipids	Diesters (Glycerol)	Diethers/ Tetraethers(Glycerol)	Diesters (Glycerol)
Membrane sterols	Scarce	Scarce	Present
Peptidoglycan	Present	None	None

Gene			
Introns	Scarce	Scarce	Present
Transcription/Translation	Present	Scarce/Present	Present
Polygenic mRNA	Present	Present	None
Poly-A termination	None	Present	Present
Ribosomes	30S/50S	30s/50S	40S/60S
Initiator AA (tRNA)	Formylmethionine	Methionine	Methionine
Metabolism			
Oxidative phosphorylation	Dependent on membrane	Dependent on membrane	Mitochondria
Photosynthesis	Dependent on membrane	Dependent on membrane	Chloroplasts (plants)
Inorganic compounds	Utilize	Utilize	Can't utilize
Anaerobic energy generation-Nonglycolytic	Occur	Occur	None
PHB (Reserve)	Occur	Occur	None
Nitrogen fixation	Yes	Yes	None
Other			
Exo/Enocytosis	None	None	Occurs
Amoeboid Movement	None	None	Occurs

The cell wall of the bacteria is made up of peptidoglycan which is a cross linked polymer. The main constituents of the peptidoglycan are repeating units of N-acetylglucosamine-N-acetlylmuramic acid. The presence of muramic acid is considered as an important characteristic of the bacteria. The lipids present on the membranes of bacteria and eukaryotes are fatty acids with straight chains mostly palimitate. The Archaea includes three different kinds of microorganisms namely thermoacidophiles, extreme halophiles and methanogens. The main difference observed in these three types of organisms is the composition of lipids. In Archaeal members, the units of glycerol are attached to ethers.

The hierarchical mode is applied to taxonomic systems for biological organisms. The system starts with Kingdom or it is also called as domain, it is followed by phylum. The phylum is also called as division and it is followed by class, order, family, genus and finally species and in some cases subspecies. By convention, the biological names are written in italics or in some cases underlined. The subspecies level can further be classified into pathovar, serovar and in some cases biotype.

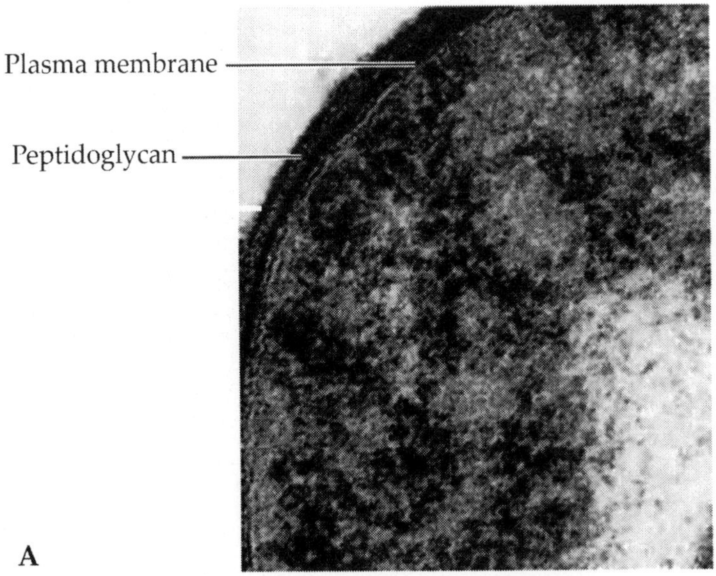

A

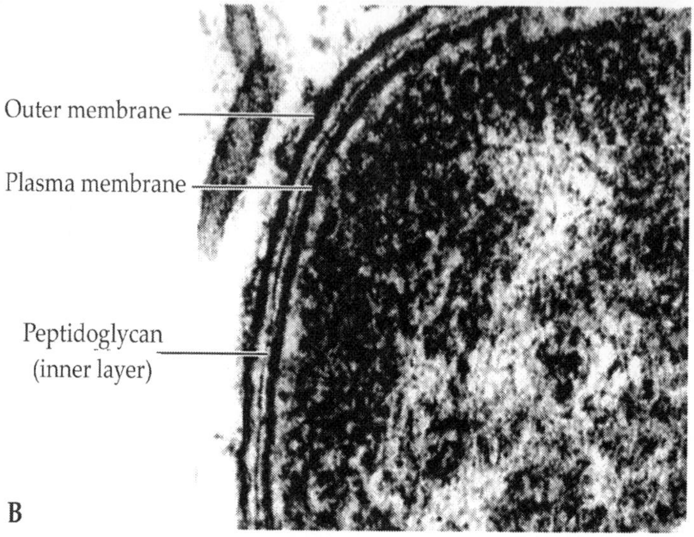

Figure 1: Source: h T. D., and Madigan, M. T. (1988). *Biology of Microorganisms*, 5th Edition, Englewood Cliffs, NJ: Prentice Hall

The rank which is designated as pathovar is also called as pathotype. This rank is applied to organisms with certain pathogenic properties. The best example is *Xanthomonas campestris pv vesicatoria*. This organism causes bacterial spot in tomato and pepper. Serovar is also called as serotype and it means the antigenic traits of certain organism. Biovar is also called as biotype and it is used to refer species or strains of certain physiological and biochemical properties.

Table 2: Taxonomic categories with ranks

Category	**Some examples**		
Domains	Archaea	Bacteria	Fungi
Phylum	*Crenarchaeota*	*Proteobacteria*	*Ascomycota*

Class	*Thermoprotei*	*Alpha-proteobacteria*	*Saccharomycetes*
Order	*Sulfolobales*	*Legionellales*	*Saccharomycetales*
Family	*Sulfolobaceae*	*Legionellaceae*	*Saccharomycetaceae*
Genus	*Sulfolobus*	*Legionella*	*Saccharomyces*
Species	*Sulfolobus acido-caldarius*	*Legionella pnue-mophila*	*Saccharomyces cerevisiae*

The ninth edition of Bergey's Manual of Sytematic Bacteriology is considered as the standard reference whilst developing computer empowered numerical taxonomy which considers several hundreds of characters which can be examined before selecting the seed microbe or the starter microorganisms for mass scale production. Much information about the microbial diversity comes from the DNA sequencing. The rRNA sequencing technology has revolutionized the field of phylogenetics. The rRNA is conserved whilst the course of evolution and hence it can be used as an identifier for many species. Lactic acid bacteria was placed distantly from rod shaped bacteria, but now after the rRNA sequencing technology arrival it is proved that the lactic acid bacteria and rod shaped bacteria are close relatives.

Bacteria are considered as one of the work horse which is being utilized in biotechnological applications. We will discuss some of the bacterial groups which have immense importance in biotechnology.

Phylums

Deinococcus-Thermus is classified into two main orders. Each order contains a single family. *Deinococcus* is highly resistant to ionizing radiation. It is a gram positive chemoheterotroph. It is successfully isolated from dust, ground meat and soil. They appear pink or bright red in colour and it is due to higher amounts of carotenoids however, this characteristic is absent in gram positive bacteria. The outer membrane does not contain lipopolysaccharide which is generally present in gram negative bacteria. *Thermus* is the genus which is gram negative and they are filamentous and rod shaped. They possess respiratory metabolism and are chemoheterotrophs and aerobic heterotrophs with thermophilic character. They are not mobile and were first isolated from hot springs in the year 1969. They can easily be isolated from hot water tanks installed in institutions and homes. *Thermus aquaticus* is being used to extract thermostable DNA polymerase also called as Taq polymerase. This enzyme is used in polymerase chain reaction (PCR) technology which is used for gene amplification. The gene responsible for this enzyme is cloned and successfully expressed in *E. coli* for mass scale production.

Proteobacteria belongs to gamma division. It constitutes enteric bacteria such as *E. coli* and *Pseudomonas* species. *Acidithiobacillus* is one of the families included in this division which constitutes several chemoautotrophic bacteria. *E. coli* is the best studied organism which harbors the intestinal tracts. It belongs to the family Enterobacteriaceae or enteric. They are gram negative bacteria and rod shaped. They possess peritrichous flagella. The key metabolism observed in these bacteria is that ATP is generated by oxidative reduction of organic compounds in air and sugar fermentation whilst anaerobic conditions. Using this bacterium, several proteins such as growth hormone and human insulin are produced via recombinant DNA technology.

Fluorescent pseudomonads are other important organism which is considered important in their applications in biotechnology. Phylogenetic studies at molecular level reveal that Pseudomonas genus is classified into alpha, beta and gamma Proteobacteria subgroups. These are gram negative bacteria and the gamma division of the genus *Proteobacteria* of genus *Pseudomonas* contains species such *as Pseudomonas syringae, Pseudomonas fluorescens, Pseudomonas putida and Pseudomonas aeruginosa.* Their flagella are located at their poles which is a significant character which demarcates them from *E. coli.* The yellowish green fluorescent compounds are released by fluorescent pseudomonads. They fail to produce ATP and generally survive as obligate aerobes. Anthropogenic substances such as halogenated aromatic compounds, octane, toluene and camphor can be degraded by the fluorescent group. So, these organisms are of extreme importance in bioremediation of eutrophicated soils. *P. syringae* is one of the plant pathogen present in fluorescent group. Pseudomonad contains outer membrane which constitutes protein complexes which enables the organism in nucleating ice crystal formation. This bacterium produces frost on leaves just below the freezing point, generally at-3°C. This causes damage to the leaf epidermis and hence it can infect the plant host. This bacterium has damaged crops and resulted in $1 billion losses every year in the U.S. In order to decrease the damages, ice'- '*P. syringae* is produced by recombinant DNA technology. This leads to gene inactivation which stops production of ice nucleating complexes. On spraying the ice'- '*P. syringae,* on leaves, the competition exists between the two species and hence the ice formation on leaves is diminished. The wild species of *P. syringae* is being utilized in production of ice crystals in sky resorts even at higher temparatures.

Xanthomonas is the genus which is a plant pathogen belonging to fluorescent pseudomonads. This genus produces yellow pigments. The species *Xanthomonas campestris* produces a polysaccharide called as Xanthan. This polysaccharide has numerous uses in oil recovery and also in the food industry.

Acidithiobacillus is a genus whose species are gram negative and rod shaped. They are small and live as obligate acidophilic mostly in pH less than 4. They generate energy by oxidizing reduced ferrous or Fe (II) and reduced sulfur compounds. Acidithiobacillus ferrooxidans or Thiobacillus ferrooxidans generally grow on Fe^{2+} enriched media. *Acidithiobacillus thiooxidans* or *Thiobacillus thiooxidans* uses or reduces sulfur compounds as their growth substitutes. In anaerobic state, the reduced sulfur compounds are oxidized by using ferric iron as acceptor of electron. *Acidithiobacillus* is a popular genus of importance due to its applications in bioleaching process of metals.

Proteobacteria, especially the alpha division has several organisms which have very interesting properties. *Rickettsia*, *Rhizobium* and A*grobacterium* are some of the important species found in one of the subgroups of alpha division. The *Rhizobium* bacteria are gram negative and are rod shaped with flagella. The *Rhizobium* species live in the soil and often attack roots of legumes. They are aerobic chemoheterotrophs. They attack roots and harbor them resulting in the formation of nodules which is a key machinery for nitrogen fixation. *Agrobacterium* is a gram-negative bacteria and it is rod shaped with flagella. They are found in soils and are considered as aerobic heterotrophs. The bacterium contains a plasmid which is large in size and it is called as Ti plasmid. This plasmid encodes different functions needed for the injection of tiny portion of DNA into plant cells. The plasmid DNA to be transferred is also referred to as T-DNA. This DNA is fused with the plant genomic DNA and initiates hormones required for production of plant tumors and galls. The gene transfer from prokaryotes to eukaryotes by natural means is observed among these species. Stable transfer of genes into the crops was successfully tried by exploiting this technique.

Zymomonas is another genus which is predominantly gram-negative rods. They possess flagella at poles. The bacterium is found predominantly in apple ciders, sugarcane extract and palm wine. They can easily survive by respiration or fermentation. The fermentation process is done by Entner-Doudoroff pathway with ethanol as end product. This species has many advantages over yeast in production of ethanol at an industrial scale.

Gluconobacter is rod shaped and rarely ellipsoidal and they are motile with flagella at polar ends. They live as obligate aerobic chemoheterotrophs. They get energy by oxidation of ethanol and converting it to acetic acid. The characteristic feature is that the acetic acid formed does not get oxidized. *Gluconobacter* is highly recommended organism as far as the production of vinegar is concerned. This organism also oxidizes glucose molecule to gluconic acid. Gluconic acid has immense commercial value.

Firmicutes is a gram-positive bacteria and it contains low DNA molecular concentration of Guanine and Cytosine content. This phylum is recognized by endospore forming capacity and the members of this phylum live as obligate anaerobes. This phylum is classified into the bacilli, the *Mollicutes* and the *Clostridia*. *Clostridium* exhibit heterogenous character. Some of the members of the *Firmicutes* phylum are *Staphylococcus,* Lactic acid bacteria and *Bacillus*. The *clostridium* is gram positive and rod shaped with flagella. They live as anaerobes and can produce endospores when favorable conditions are absent. These organisms are of biotechnological value because the end products are acetone, butanol, acetylmethylcarbinol and ethanol. *Clostridium acetobutylicum* produced twelve tons of acetone by utilizing hundred tons of molasses. The technology of genetic engineering has improved the fermentation by *clostridium* for acetone production.

The group comprising *Lactobacillus*, *Staphylococcus* and *bacillus* forms a cluster. They live as facultative anaerobes. The *Lactobacillus*,

Leuconostoc and *Pediococcus* form a group which can be named as lactic acid bacteria. They get energy by fermenting glucose resulting in lactic acid. Carbon dioxide, ethanol and lactic acid are the end products in *Leuconostoc*. Flagella is absent in most of them. Round shaped members are predominant in *Pediococcus* and *Leuconostoc* whereas rod shaped members are seen in *Lactobacillus*. Appreciable growth is observed in these members especially when oxygen is limited and if the media is enriched with vitamins, pyrimidines, purines, peptides and sugars. Several strains of bacteria are being used to produce yogurt, buttermilk, butter and cheeses.

Streptococcus is circular in shape and they generate ATP by transforming glucose into two lactic acid molecules. *Streptococcus cremoris* is used in manufacturing Cheddar and Gouda which are hard pressed cheeses. The bacteria also produce Camembert which is a soft ripened cheese. Lactobacilli and streptococci are used by the meat and dairy industries as starter cultures.

Bacillus is chemoheterotrophic in nature and they are aerobic. They live in soil mostly and can growth with an exponential rate if media is supplied. Strains of bacillus are thermophilic in nature and they are capable to grow even at 75°C. The bacterium produces several hydrolytic enzymes which have the capacity to cleave lipids, polysaccharides, nucleic acids, and proteins. These enzymes are required to be produced on a large scale. The proteolytic enzymes form a main ingredient of detergents used in dry laundries. Biological insecticides protect crop and one of them is *Bacillus thuringiensis*. Polymyxin and Bacitracin are the antibiotics produced from bacillus strains of Bacillus polymyxa and Bacillus subtilis.

Staphylococcus inhabits human and animal skin. They are non-motile and spherical in shape. They do not produce endospores. They exhibit outstanding tolerance towards salt and drying. This bacterium can inhabit in air indoors, dust, animal feeds, poultry and meat. The protein A extracted from this bacterium is used in analytic processes and purification of proteins.

Actinobacteria is a gram-positive bacteria having higher concentration of GC content in its DNA. They can form filaments and can irregularly divide. They are rod shaped and are without flagella. Cellulomonas and Arthrobacter form one cluster. Coryneabacterium, Mycobacterium and Nocardia from another cluster. The second cluster is basically soil microorganisms and they live in aerobic conditions. Their cell wall is comprised of mycolic acids which are long chained fatty acids. *Cellulomonas* can decompose cellulose. *Corynebacterium glutamicum* can produce glutamic acid and this bacterium has been used extensively for the production of amino acids on a large scale.

Streptomyces produce filaments called hyphae. They produce a network of filaments called as mycelia. On maturing, the filaments projects above the surface. The hypha which is aerial produces cross walls and each cell form into spores which are called as conidia. It can break chitin, pectin, starch and proteins. Streptomycin is extracted from *Streptomyces griseus*. Antibiotics such as gentamicin, neomycin, erythromycin and tetracycline are being produced on a large scale by using *Streptomyces*.

FUNGI

They are eukaryotes and have the ability to produce spores through asexual and sexual reproduction. The hyphae show apical growth. They cannot participate in the process of photosynthesis. Many of them are parasites and few live as symbionts. They have the capacity to absorb nutrients through their cell walls. Several enzymes are produced by these organisms to degrade substances with high molecular weight. The cell walls are made up of polysaccharides and are rigid.Fungi can be classified into five groups namely Chytridiomycota, Glomeromycota, Zygomycota, Ascomycota and Basidiomycota.

Chytrids of Chytridiomycota contains around a thousand species. The cell walls are made up of beta1-4 N-acetylglucosamine which is a chitin. The reproductive cells contain flagella which give them the ability to swim, the other members of fungi lack flagella, they grow in marine and freshwaters and several members are saprophytic. *Rhizophlyctis rosea* decompose cellulose. Glomeromycota constitute members of *Arbuscular mycorrhiza* (AM). They are asexual organisms and often symbiotic. The AM represents their attachment to the plant roots. They make the soil fertile. In one cubic millimeter of soil, the AM can produce mycelia of about 20,000 kilometers. The mycelium can absorb inorganic nutrients such as phosphate from the soil substratum. Greater economic importance can be achieved by AM fungi. These fungi increase biodiversity and it can be also used against fungal pathogens and nematodes.

Zygomycota generally produces zygospores and they are asexual spores. The zygospores are formed in a sporocarp. Cross walls are absent in the mycelium. Non-acetylated glucosamine or chitosan and chitin are the main ingredients of cell wall. *Rhizopus* and *Mucor* are important members of Zygomycota which are saprophytic in nature. *Rhizopus nigricans* is extensively used for citric acid production.

Ascomycota is an important division comprising of around 30000 species. The character that defines the Ascomycota is the ascus and ascospores are formed within the ascus. Ascomycota exist as single cells as seen in yeast or segmented filamentous structures.. Sexual reproduction results in spores and they are present in ascus. *Saccharomyces* or brewer's or baker's yeast and *Neurospora* or bread mold are economically important species of Ascomycota. *Saccharomyces cerevisiae* multiply by the process of budding.and isis the main yeast used in production of wines,beers and also bread

Basidiomycetes produce basidium and the cell walls are made up of chitin and glucans. *Ustilago maydis, Puccinia graminis* and *Serpula lacrymans* are some of the examples of this sub division. Agaricus is one of the commercial product in this subdivision.

Due to the noticeable fungi in only asexual stage, they were placed in deuteromycetes. Due to asexual reproduction, conidia are produced and sexual reproduction is completely absent. Like ascomycetes, basidiomycetes the mycelium is septate and unicellular. The fungal members in this group are also called as "fungi imperfecti". Species of *Pencillium* and *Aspergillus* have a big economic importance. *Aspergillus niger* produces gluconic acids and citric acid on a commercial scale. *Aspergillus oryzae* is being used to produce amylolytic enzymes and proteolytic enzymes. *Aspergillus flavus* produces aflatoxin B_1. *Penicillium griseofulvum* is used to produce griseofulvin and it is used as anti-fungal agent to treat infections caused by fungi on nails and skin. *Penicillium roqueforti* and *Penicillium camemberti* are used to produce cheese.. Yeast grows even in acidic conditions and some are resistant to antibiotics.

DNA sequencing and PCR are the latest techniques which provide us sufficient knowledge about the industrially viable organisms. Protein synthesis is considered as one of the important mechanisms for the growth of an organism. Transcription allows the passage of information from DNA molecule to messenger RNA (mRNA) The mRNA formed in transcription is transported out of the nucleus, into the cytoplasm, to the ribosome, then mRNA directs protein synthesis. This process is called translation and also needs the assistance of transfer RNA (tRNA). (Translation). These two processes are named the Central Dogma of Molecular Biology (DNA-RNA-proteins). In retroviruses, the synthesis of DNAfrom an RNA template is made via a process called as reverse transcription. The retroviruses are used in the field of biotechnology for various purposes.

Transcription

In transcription, RNA polymerase is used to denature DNA which is to be transcribed later. A single strand of the DNA is transcribed into RNA. This strand is called as sense strand and the other strand which is not participating the transcription process readily is known as anti-sense strand. Once mRNA is formed, it gets migrated into the cytoplasm and ribosomal machinery.

Translation

In every cell, the organelles where the protein gets synthesized are the ribosomes. They comprise of one-third protein and two-thirds of ribosomal RNA. The ribosomes comprise of two sub-units one smaller and the other larger. The smaller sub-unit is named as 30S and 40S sub-unit and the larger sub-unit is called as 50S and 60S sub-unit for prokaryotic and eukaryotic cells, respectively the units are Svedberg unit which is dependent on ultra-centrifuge readings. The smaller sub-unit contains a site for mRNA binding. Two binding sites are present on the larger subunit which contains two binding sites for transfer RNA. The mRNA is considered as the blueprint for the synthesis of protein. Translation is nothing but the joining of amino acids to form proteins. The bases present on the mRNA act as codons. 64 types of codons are present in the process of translation, however, 61 codons codes for aminoacids which later form proteins. The other three codons are considered as stop codons and they are involved in halting the process of protein synthesis. The starting codon is AUG and it also codes for amino acid Methionine.

The promoters are the DNA sequences which are considered as the start signals for the mRNA transcription. The mRNA is generally made up of 10,000 nucleotides or the number may vary. The translocation process occurs in the ribosomes. The ribosomes migrate or move onto the messenger RNA and keep the amino acids together to form proteins. The genetic code is given below.

The Transfer RNA generally collects the amino acids to the site of protein synthesis and it helps in extension of amino acids which leads to formation of proteins. The tRNA molecule is clover leaf shaped structure. The attachment of amino acids to the tRNA is controlled by the enzyme called as aminoacyl-tRNA synthetases. The amino acids get attached to tRNA and the energy required is supplied by the cleavage of ATP into AMP. Once the synthesis of protein is complete, the folding commences to expose the functional chemical groups which infer certain peculiar characters to the protein. The folding can be secondary, tertiary and quadruple. Due to the coiling of proteins, a three-dimensional structure is generated and it is of immense biological value.

The polymerase chain reaction is used to make multiple copies of DNA from very minute amounts of DNA. This process as invented by Kary B. Mullis in the year 1985. The requirements of the polymerase chain reaction or the PCR are

- I DNA or RNA samples which are to be multiplied
- 2 primers (Reverse and Forward)
- Deoxynucleotide triphosphates (dNTPs) which are the monomeric substrates for the polymerization reaction. It is a mixture of the four monomeric units, dATP, dTTP, dCTP and dGTP.Taq polymerase which is extracted from *Thermus aquaticus* andis a thermostable enzyme.

The primer is the small nucleotide segment and it is designed in such a way that it is complementary to the DNA to be multiplied. The primers attach to the DNA template through the process called as annealing. This supply an initiation site for the DNA elongation which is to be formed afresh. The polymerase chain reaction occurs in three steps. They are

- Denaturation step which is carried at 94-96^{0}C
- Annealing step which is carried at 54-60^{0}C

- Extension which is carried at 72^{0}C

In the first step, DNA is subjected to heating(94-96^{0}C) and this causes the separation of the the two strands of DNA which is also known as denaturation. In the anneling step, the solution temperature is cooled from 94-96°C to 50-60°C and the primers (short sequences complementary to the target sequence of DNA) previously added to the PCR mixture bind to the single strand of DNA.. In the extention step, the enzyme Taq polymerasestart off by binding to the primers and then extend the primers by adding new nucleotides to the 3' end, using the single stranded DNA as a template.

The polymerase chain reaction is much efficient and can be performed easily. Some of the important applications of the polymerase chain reaction are

- It produces significant amount of DNA which can be used in the process of genetic engineering and also in sequencing.
- This technique helps us in determining the taxonomic position of certain organism which can be used for biotechnological application with greater certainty.
- This technique can also be used to detect the casual organism of contamination and it helps us to eliminate the contaminants

Microarrays

Microarray technology rests on the ability to deposit many different DNA sequences on a small surface, usually a glass slide (often referred to as a chip, DNA chips, biochips or microchips). The DNA sequences are immobilized onto a certain location of the solid substrate. The different DNA fragments are arranged in rows and columns such that the identity of each fragment is known through its location on the array. The support is generally nylon membranes or silicon chips or microscopic slides made up of glass. The DNA can be directly synthesized or spotted onto the support. . This process depends on the process called as hybridization probing. The microarrays contain single stranded sequences which are labeled with a fluorophore. The unknown sample is allowed to get hybridized with the labeled and fixed DNA on the supports. The patterns are studied and compared with the standard. The DNA strand which is immobilized on the array is called as probe and the DNA strand which is in the solution is known as target.

Some of the advantages of the Microarrays are:

- This technique exhibits high through-put i.e. many elements can be deposited in a tiny area and this helps us to analyze even at the genomic levels. A large number of samples can be monitored simultaneously within an experiment.
- This technique exhibits higher rates of sensitivity and reproducibility since the oligonucleotide sequences printed on the chip that is designed for a specific gene is perfectly complementary to the target gene sequence.
- Simultaneous observation of more biological samples is possible through this technology via differential display.
- The interference by the background contaminants in the experiment is low. Since the non-specific binding rate on the solid surface is low, the elimination of the fluorescent and organic contaminants is quick.
- This technology is automated and it makes the process economical when compared with other nucleic acid based technologies.

DNA sequencing

DNA sequencing is nothing but the determination of exact nucleotide sequence in the given DNA sample. There are two methods which are available for DNA sequencing. They are Maxim and Gilbert method and the Sanger's method.Sanger's method is widely used. Sanger's method is known as dideoxy method and it is also called as enzymic method of DNA sequencing. The nucleotides used in this method don not have the 3'-hydroxyl group and so it is called as dideoxy method. Dideoxynuclotide triphosphates or the ddNTP are used in this method. Due to absence of 3'-hydroxyl group, the elongation of DNA chain cannot be extended further, so that termination occurs specifically at positions where dT should be incorporated. Due to this property, this method is also called as chain termination method. Single stranded DNA which is to be sequenced is added to the primer along with DNA polymerase I and higher amounts of nucleotide triphosphates. Approximately 5% of dideoxynuclotides are added and each molecule is dyed with different fluorescent colours. The starting point and direction of chain elongation is decided by the primer. The synthesis of DNA starts with the primer and the chain ceases to elongate once ddNTP gets injected into the sequence instead of dNTP. Therefore, the fragments formed will be of different lengths. The four nucleotides run separatelywith ddNTP. The mixture with ddTTP will produce fragments with Thymine terminals and the mixture with ddCTP results in Cytosine terminals and so on. The fluorescent DNA strands are separated from the template and polyacrilamide gel electrophoresis is conducted to separate as per the lengths. Four lanes are produced after the gel is read. Since ddNTP's are having different colours, the scanner will have the power to read and record the colour of each band. Sanger's method is eing applied to the DNA fragments ranging from 700-800 nucleotides.

Bioinformatics

The utilization of computer systems to simulate, predict, analyze, retrieve, compare, and store the genetic molecular structure and composition is called as Bioinformatics. These areas mainly simulate proteins, DNA and RNA. The main research in the bioinformatics revolve around modeling of evolution, protein-protein interactions, prediction of gene expression, prediction of protein structure, structural alignment of protein, genome assembly, gene finding and sequence alignment. Bioinformatics uses data obtained from high throughput bio-techniques and applies mathematical logics. The enhanced computational capacity allowed researchers to obtain data from microarray chips, gel electrophoresis and PCR and hence capability of doing proteomic and genomic analysis. The area of bioinformatics is being used in

- Comparative genomics and sequencing, in gene identification and to understandthe genome function .
- Protein identification and simulation of regulatory and metabolic pathways (proteomic).

- Simulation and visualization of the cell and modeling the cell behavior.
- It is also used in drug discovery and in production of antimicrobial chemicals.

Some of the applications of bioinformatics in the biotechnology field are pair-wise comparison of genomes, 3D modeling of structures, identification of structure and function of gene and genomic sequencing.

The microorganisms secrete several low molecular weight chemicals or compounds which are called as secondary metabolites. Some of the secondary metabolites extracted from bacteria and fungi are Actinomycin, Bleomycin, Griseofulvin, Rifamycin, Chloramphenicol, Tetracyclin, Lincomycin, Ertythromycin, Cycloheximide. Puromycin, Fusidic acid, Cycloserine, Bacitracin, Penicillin, Cephalosporin, Vancomycin, Telcoplanin, Polymyxin, Amphotericin, Gramicidin, Monensin, Avermectins, Clavulanic acid, Kasugamycin, Polyoxins, Nikkomycin, Spinosins, Bialaphos, Cyclosporin A, Fk-506, Rapamycin, Doxorubicin, Ergot alkaloids, Lovastatin or Mevinolin, Gibberellins and Zearalenone.

SUMMARY

Prokaryotes and Fungi comprise a large number of organisms which have varied metabolic products, sources of energy, carbon source and the ability to infect other species. The classification of fungi is based on the molecular analysis. The kingdom fungi are classified into five divisions. The most exploited fungi are Yeasts. Very small number of bacteria and fungi are exploited till date.

REFERENCES

1. The Authors. (2007) Crystal Ball – 2007. Environmental Microbiology, 9, 1–11.
2. Ingraham, J. L., and Ingraham, C. A. (2004). Introduction to Microbiology: A Case
3. History Approach, 3rd Edition, Pacific Grove, CA: Brooks/Cole.
4. Gest, H. (2003). Microbes: An Invisible Universe, Revised Edition, Washington, DC:

5. ASM Press.
6. Dyer, B. D. (2003). A Field Guide to Bacteria, Ithaca: Cornell University Press.
7. Madigan, M.T., Martinko, J. M., andParker, J. (2003). Brock Biology of Microorganisms, 10th Edition, Upper Saddle River, NJ: Prentice Hall.
8. Lengeler, J.W., Drews, G., and Schlegel, H. G. (ed.) (1999). Biology of the Prokaryotes,
9. Malden, MA: Thieme. Microbiology Online Resources http://www.nature.com/nrmicro/info/links.html.
10. Garrity, G. M. (ed.) (2001). Bergey's Manual of Systematic Bacteriology, 2nd Edition,
11. New York: Springer-Verlag.
12. Amann, R., and Schleifer, K.-H. (2001). Nucleic acid probes and their application in environmental microbiology. In Bergey's Manual of Systematic Bacteriology, 2nd Edition, Volume 1, The Archaea and the Deeply Branching and Phototrophic Bacteria,G.M. Garrity (ed.), pp. 67–82, New York: Springer-Verlag.
13. Zhang, Z., Willson, R. C., and Fox, G. E. (2002). Identification of characteristic oligonucleotides in the bacterial 16S ribosomal RNA sequence dataset. Bioinformatics, 18, 244–250.
14. Loy, A., et al. (2002). Oligonucleotide microarray for 16SrRNAgene-based detection of all recognized lineages of sulfate-reducing prokaryotes in the environment. Appl. Environ. Microbiol., 68, 5064–5081.
15. Sebat, J. L., Colwell, F. S., and Crawford, R. L. (2003). Metagenomic profiling: microarray analysis of an environmental genomic library. Applied Environmental Microbiology, 69, 4927–4934.
16. Burggraf S., Mayer, T., Barns, S. M., Rossnagel, P., and Stetter, K. O. (1995). Isolation of a hyperthermophilic archaeum predicted by in situ RNA analysis. Nature, 376, 57–58.
17. Koonin, E. V., Makarova, K. S., and Aravind, L. (2001). Horizontal gene transfer in prokaryotes: quantification and classification. Annual Review of Microbiology, 55, 709–742.

18. Achenbach, L. A., and Coates, J.D. (2000). Disparity between bacterial phylogeny and physiology. ASM News, 66, 714–715.
19. Brunner, R. (1975) Biotechnologie in Osterreich. 1. Arbeitstagung, Graz.
20. Davies, J. (1982) Nature 299, 493.
21. Ettler, P. (1991) Folia Microbiol. 36, 493.
22. Ginjaar, L. (1980) Biotechnol. Lett. 2, 153.
23. Hartmeier, W. (1985) Trends Biotechnol. 3, 149.
24. Hockenhull, D. J. D. (1975) Adv. Appl. Microbiol. 19, 187.
25. Jacob, H. E. (1954) 6 000 Jahre Brot. Hamburg.
26. Klibanov, A. M., Samokhin, G. P., Martinek, K., Berezin, I. V. (1977) Biotechnol. Bioeng. 19, 1351.
27. Kramer, S. N. (1965) History Begins in Sumeria (In Czech), Prague.
28. Ktienzi, M., Assi, F., Chmiel, A., Collins, C. H., Donikian, M., Dominguez, J. B., Financzek, L.,
29. Fogarty, E. M., Frommer, W., Hasko, F., Howland, J., Houwink, E. W., Mahler, J. L., Sandkvist,
30. A., Sargeant, K., Sloover, C., Tu~'nenburg Mu~s, G. (1985) Appl. Microbiol. Biotechnol. 21, 1.
31. Lafferty, R. M. (1975) Biotechnologie in Osterreich. 1. Arbeitstagung, Graz.
32. M/tlek, I. (1976) Physiological State of Continuously Grown Microbial Cultures. In: Continuous
33. Culture 6: Applications and New Fields, Chichester.
34. Marcus, I. (1975) Adv. Appl. Microbiol. 19, 77.
35. Moser, A. (1992) Acta Biotechnol. 12, 69-78.
36. Pontecorvo, G. (1976) Presidential Address. In: 2nd Internatl. Symp. Genet. Industrial Microrganisms,
37. London.
38. Pridham, T. G., Hesseltine, C. W. (1975) Adv. Appl. Microbiol. 19, 1.
39. Ri6ica, J., Votruba, J. (1982) Overproduction of Microbial Products. FEMS Symp. No. 13, London.

40. Sikyta, B., Prokop, A., Novfik, M. (1973) Advances in Microbial Engineering, Part 1. Biotechnol. Bioeng. Symp. 4, New York.
41. Esser, K. (ed.) (2004). The Mycota: A Comprehensive Treatise on Fungi as Experimental Systems for Basic and Applied Research, 2nd Edition, Berlin; New York: Springer-Verlag.
42. Watling, R. (2003). Fungi, London: Natural HistoryMuseum.
43. Dighton, J. (2003). Fungi in Ecosystem Processes, New York: M. Dekker.
44. Alexopoulos, C. J., Mims, C. W., and Blackwell, M. (1996). Introductory Mycology, 4th Edition, New York: JohnWiley.
45. Sch¨ussler, A., Schwarzott, D., and Walker, C. (2001). A new fungal phylum, the Glomeromycota: phylogeny and evolution. Mycological Research, 105, 1413–1421.
46. Guarro, J., Gen´e, J., and Stchigel, A. M. (1999). Developments in fungal taxonomy. Clinical Microbiology Reviews, 12, 454–500.
47. Berbee, M. L., and Taylor, J.W. (1993). Dating the evolutionary radiations of the true fungi. Canadian Journal of Botany, 71, 1114–1127.
48. Pennisi, E. (2004). The secret life of fungi. Science, 304, 1620–1622.
49. Wardle, D. A., Bardgett, R. D., Klironomos, J. N., Set¨ala, H., van der Putten, W. H., and Wall, D. H. (2004). Ecological linkages between aboveground and belowground
50. biota. Science, 304, 1629–1633.
51. Tree of Life Web Project http://tolweb.org/tree?group=Fungi&cont group=Eukaryotes#TOC6.

CHAPTER 2

MICROBIAL GROWTH KINETICS

INTRODUCTION

The knowledge of microbiological process is mandatory and it is gained by studying kinetics related with production rates of cell along with the metabolites and the factors which control them. Advances in enzyme studies, product formation, and growth regulation lead to understand the microbial growth kinetics with more certainty. These advancements in enzyme systems research helps to regulate the course in industrial production at a mass scale and also helps in generalization. This chapter focuses on the microbial growth kinetics involved in case of batch cultures and it explains the importance of various factors affecting the processes at an industrial scale.

Growth of microorganisms and reproduction

Batch culture is classified as one of the basic methods used in cultivation of microorganisms. The submerged type of cultivation which generally occurs in liquid medias is considered as an important composition of batch culture. A pre-determined concentration of sterile medium is used for a given species of microorganisms that are to be cultured in specific set of conditions such as aeration, temperature, etc. The growth and proliferation of the microorganisms cease due to accumulation of toxic chemicals and exhaustion of essential nutrients. The variations in rates of reproduction and growth decide the microbial population development.

Multiplication and growth curve

A characteristic curve is imminent if one plots the cell mass concentration or cell number or logarithms against time. If the cell mass concentration is plotted against time, then it is called as growth curve. If the cell number is plotted against time, then it is called multiplication curve. The multiplication curve and the growth curve may not be identical. Some of the phases which are characteristic while studying multiplication curve are

1. The initial growth phase, during which cell number remains rather constant prior to rapid growth is called as lag phaseIn this phase the cells which are inoculated demands some time to get adapted to the medium conditions.
2. As the time progresses, the number of cells increases rapidly. This phase is called as exponential phase.
3. The rate of multiplication starts declining due to limiting factors such as exhaustion of nutrients and release of toxic chemicals. The rate of multiplication of cells is in equilibrium to the cell death rate. This phase is called as stationary phase.
4. The death phase is classified into three phases. They are
    a. Accelerated death phase
    b. Exponential death phase
    c. Death declination phase

The lag phase is normally shorter (if the conditions are favourable) than the exponential phase. The linear relationship does exist between the cell weight and the cell number. During the exponential phase, the multiplication and growth rates are at the maximum. The rate at which the cell components such as enzymes, DNA, RNA etc. is almost identical. This period is too short and it is called as balanced growth. The stationary phase is the longest phase. Afterwards, there will be declination of biomass concentration due to cell lysis. The difference observed in these two curves is mainly due to varies cell sizes and their composition because of degradation and accumulation of cell debris and associated toxins whilst their culture. Except for the period of balanced growth, the changes are evident in growth and multiplication rates. The growth kinetics can be more complex and the there might be phases missing. When microorganisms are transferred into new media's, they exhibit constant concentration for a limited period of time. It may take more hours or sometimes it takes very short time. **Lag phase** The key factors which affect the period of the Lag phase are the components of the medium where the microorganism were inoculated , the strain of the microorganism, age, concentration of microorganisms and the physical factors which includes pH, redox potential, temperature and the genetic properties. The period of the this phase is the time taken from the point of their inoculation and the time when the microorganisms starts to grow exponentially. Monad in the year 1949 explained that the period between parallel ideal line and the experimental line as lag phase. The first 2 phases can be described as

a) The rate of growth exceeds the rate of multiplication. The activity of the enzyme can be studied by observing the rate of oxygen consumed and the amount of heat evolved or the carbon dioxide produced. This is often referred to as biomass unit.

b) The nature of medium decides the period of lagphase. The microorganisms which are taken from exponential phase generally exhibit negligible lag phase but the microorganisms taken from stationary phase shows extended acceleration and lag phases.

c) The other phases of growth are not affected by lag phase in general.

d) If inoculum taken is larger, then there would be negligible lag phase.

e) In lag phase, the cells are more sensitive to environmental changes such as chemicals, pH and temperature and also do not agglutinate when subjected to electrophoresis.

Exponential phase

The lag phase and acceleration phase leads to the exponential phase. In the exponential phase, maximal growth rate of microorganisms is observed. The cells replicate at the same rate and do not change with time. The rate of multiplication is maximum and it is due to biophysical and biochemical processes.

Stationary phase

This phase is characterized by the maximum microbial population for a specific period. The concentration of microorganisms depends on various factors which occur simultaneously such as the oxygen concentration, trace elements, nitrogen, concentration of energy source at initial stages and also the methods used in regulation of pH especially organic acid or inorganic acid. It the essential nutrient concentration is less, then the growth in total is directly proportional to the concentration. The cells in this phase produce reserve substances and they prepare to sporulate.

Death phase

There are no hard and fast rules to determine the cell death. The death of a microbe can be fast or in some cases it can be slow. The death of the microbe can be associated with autolysis. In batch culture, the variations observed in the cell concentration are due to physiological state of the cell at different times i.e. the physiological and biochemical properties (Malek, 1976). Herbert (1961) made some observations while studying these changes. When the microorganisms are inoculated into fresh medium, the mass of the cell increases rapidly and this increase equals to the rise of RNA content. In the exponential phase, the individual cell mass and the RNA concentration per cell weight remains constant. When a specified essential nutrient is reduced in the medium, the cells mass gradually decreases.

Escherichia Coli exhibit more sensitivity towards UV light. The sensitivity rises in the lag phase and extends till the exponential phase, soon declines. These changes are due to deamination activity, carbondioxide production, oxygen consumption and changes in metabolic activity. In batch cultures, certain microorganisms exhibit remarkable changes in their cell morphology and also growth cycles. *Azotobacter chroococcus* exists in several forms and hence the culture seems to be containing various microbial species. At the start of the culture cultivation, *Azotobacter* cells appear like motile rods and the ends are rounded and they can be single or double. As the time progresses, shortening of rods is seen and the cells become coccus shaped of diplococcus shaped and they appear as grainy structures and they aggregate mostly. These species form cysts. *Streptomyces rimosus* exhibit similar pattern and it is also useful in production of oxytetracycline. Some of the members of fungi and Actinomycetes exhibit the same patterns.

Mathematics of microbial kinetics

The growth and proliferation of the microorganism's exhibit complexity in their processes. When applying mathematical principles, it should be kept in mind that there exists complete homogeneity in the culture i.e. Microbial genetic stability and physiological stability and of course the environment in which they live in.

In standard conditions, the biomass concentration with a short time interval can be directly proportional to the biomass quantity at the given time. It can be expressed as

$$dX = \mu X dt$$

where dX is the biomass concentration

dt is the time interval

X is the biomass quantity.

This equation is followed by

$$\frac{dX}{dt} = \mu X$$

Where dX/dt denotes the rate of population growth

μ represents the rate of growth of biomass unit

$$\frac{1}{x}\frac{dX}{dt} = \mu$$

It is called as specific growth rate. It is shown in reciprocal time units.

In batch cultivation, the specific growth rate ranges from zero to the maximum value. The maximum value is represented by μ_{max}. It depends on the cultivation conditions which are both physical and chemical.

If μ is taken as constant then the equation would be

$$\ln X = \ln X_0 + \mu t$$

where X_0 is the concentration of biomass at t.

The ln X is dependent on time and it gives a straight line with slope as μ. On logarithmic transformation, we get

$$\text{Log } X = \frac{\mu t}{2.30} + \log X_0$$

- It shows that

$$\ln \frac{X}{X0} = \mu t$$

$$\text{So } X = X_0 e^{\mu t}$$

This law is known as logarithmic or exponential law. The main character which controls the growth rate principally is specific growth rate. The relation between the doubling time of biomass t_d and the specific growth rate μ can be given by

$$X = 2X_0 \; ; \; t = t_d$$

$$\text{So } t_d = \frac{ln2}{\mu} = \frac{0.693}{\mu}$$

Degree of multiplication

It is the ratio X/X_0 and is generally equal to $e^{\mu t}$. If we a assume that the biomass increases by n generations then

$$\frac{X}{X0} = 2^n$$

So $n = 3.32 \log \frac{X}{X0}$

If the inoculation is 10% in v/v then the final biomass quantity, then the value of n = 3.32

The growth kinetics observed in filamentous microorganisms is complicated than the unicellular microorganisms. The filamentous microorganisms comprises of fungi, Actinomycetes and bacteria. Based on the type of strain used and the cultivation parameters followed, the filamentous microorganism's results in two modes of growth. They are homogenous and pellet.

The Actinomycetes shows filamentous growth with a specific growth rate. They form filaments which are polynuclear and aseptate. Due to the shear forces and the cavitations the filaments will be broken into tiny fragments. The ends of filaments are also called as apical cells and the length of the hyphae shows growth at constant rate. The fungal form grows into long hyphae and they differ from Actinomycetes in various characteristics. Bergter (1978) explained the growth in fungal members through

$$\frac{dL}{dt} = \alpha L_a$$

Where L is the length of hyphae and L_a is the number of the apical cells and filament ends.

The increase in hyphae length or filament is directly proportional to number of apical cells and filament ends.

The mechanical forces play a crucial role in pellet formation. These forces cleave the filaments and results in pellets. The nature of the medium selected also decides the rate of filament formation. The pellet surface is fed with oxygen and other nutrients however the growth cannot be reduced due to formation of toxic chemicals released. The centre of the pellet shows reduced filament and hyphae growth rate due to limited supply of oxygen and nutrients. It is difficult to understand the kinetics involved in growth of the pellet. It is assumed that there might be exponential growth. It is given as

$$\frac{dX}{dt} = \mu X_p$$

X_p is the biomass present on the surface of the pellet.

The radius of the pellet is directly proportional to the thickness and it is represented as

$$\frac{dr}{dt} = \mu X_p$$

With integration, it gives

$$r = \mu wt + r_0$$

If the density of all the available points is taken as d

$$X = \frac{4}{3} r^3 \pi d$$

After making necessary modifications and inclusions, we get

$$X^{\frac{1}{3}} = \frac{4\pi d}{3} \mu wt + X_0^{\frac{1}{3}}$$

Different diameters are observed in the pellets as described by Pirt (1966, 1975) with Righelato (1975).

Growth kinetics of microbial colonies

The spore or the vegetative cell is inoculated on the solid medium and the spore reproduces post germination and till the nutrients gets depleted or due to release of toxic chemicals. The cell mass or the colony at certain time can be expressed as

$$\frac{dM}{dt} = \mu M$$

On integration, it gives

$$\ln M = \ln M_0 + \mu t$$

M_0 is the original cell mass.

The colony produced may be circular with height h and radius r, p is the mass of the cell / unit volume.

then M = $\pi r^2 hp$

On integration, it gives

$$\ln r = \frac{\mu}{2}t + \ln r_0$$

r_0 denotes the radius of the original cell at time t.

lnr Vs t are linearly related to growth rate i.e. specific growth rate.

Pirt (1967) proposed linear growth model. According to this model, the cells if supplied nutrients can grow exponentially and the colonies which are grown on the dish needs to be small. As the size of the colonies increase, the diffusion rate of the nutrients into the cells reduces and as the time progresses, the growth stops.

The mass of the total colony can be given by

$$\frac{dM}{dt} = \mu M_g$$

M_g is the mass of cells which are present in the growing zone.

$$\frac{d_r}{d_t} = \frac{\text{Ä}a}{h}\mu$$

$$\text{So } r = \frac{\text{Ä}a}{h}\mu t + r_0$$

The formation of product

The formation of products can be viewed as

- It can be a process which involves regulatory mechanisms and also devoid of regulatory mechanisms (Bergter, 1983).
- I can be seen as the primary or the secondary metabolites which are generally produced during idiophase or trophophase (Bu'Lock, 1965)
- It can be extracellular or intracellular product.
- It can be viewed according to the rate of formation of product

and the growth rate (Maxon, 1955).

- It can be based on stoichiometry of the product formation.
- It is based on the rate of consumption of substrate and the rate of formation of product (Garden, 1959).
- It is also based on the complexity occurring in chemical reactions which leads to formation of products (Deindoerfer, 1960).

The rate of formation of product is controlled by the factor same as the growth rate

$$\frac{dP}{dt} = k_p \frac{dX}{dt}$$

But dX/dt equals to μX

So $$\frac{dP}{dt}\frac{1}{X} = k_p \mu$$

It is inferred that the ratio of the rate of formation of product to the specific growth rate is constant. This is called as growth associated product formation.

The rate of product formation generally depends on the microbial concentration and the physiological state of the microbe at that given point of time. It is not dependent on the medium concentration and the components present in it under specific conditions.

So $$\frac{dp}{dt} = \mu X$$

The microbial growth in a batch culture, P is the product synthesized at time interval dt

The microbial growth can be defined as the systematic increase in all the components in presence of medium and environment. There can be four types of growth observed in microorganisms. They are

- Growth by binary fission
- Growth by budding (Yeast)
- Growth by chain elongation (Fungi)
- Growth in host cells (intracellular-Viruses)

The cell number and growth of cell mass can quantitatively in terms of cell number doubling in a point of time. The techniques use to measure the growth can be by direct counting or indirect cell counting. The dry weight of the cell in a specific volume of the media or culture is one of the methods used to measure the cell growth. In this process, the cells need to be isolated from the medium and later dried. Spectrophotometer is used in estimating the absorbance of the suspensions where microbial cells are present. At particular wave length, the absorbance is dominant and it is shows proportionality with the cell concentration. If we plot the cell concentration against absorbance, we can get a standard curve. Through this curve, the cell concentration of the given sample is measured. Direct counting using microscopes and counting chamber can have certain limitations since the demarcation between live and dead cells is not clear. The electronic counting chambers can be used to quantify and also detect the size distribution of the microbial cells. This method is generally used in measuring eukaryotic cells such as erythrocytes. Plate counts are another technique which is a kind of indirect viable cell counting. In this technique culture dilutions and plating on nutrient agar medium is common. Each unit which is viable has the capacity to form a colony. These colonies can be taken as colony forming units (CFU). The CFU number is relevant to the viable count of the sample. The measurement of turbidity is a swift and non-destructive procedure to count number of bacteria in broth or clear liquid media. This turbidity method cannot measure the density of the cells which are less i.e. 10^7 cells per milliliter. The biochemical activity such as production of ATP, carbondioxide, oxygen uptake can also be detected and measured. The generation time can be defined as the growth rates of bacteria which are usually in exponential phase under standard temperature, pH and culture medium. The bacterial generation time generally range between 12 minutes and one day. *Rhizobium* bacteria exhibits extended generation times. *Mycobacterium tuberculosis* shows extended generation times and it helps in projecting the virulence which is advantageous to the microorganisms.

If the bacterial growth is due to binary fission, then increase in population of bacteria shows geometric progression. The generation time is given by

$$G = t/n$$

G denotes the generation time

N denotes the number of generations

t is the time in minutes per hour

If the growth is by binary fission, then

$b = B X 2^n$

b denotes the number of bacteria after certain time interval

B denotes the number of bacteria at initial time interval

n denotes number of generations

It can also be given as

$\log b = \log B + n\log_2$

The cell growth can be considered as the transformation of substrate into the products and cell mass. The growth of microbe's especially heterotrophic and aerobic species by computing the following equation which is also a stochiometric equation.

Substarte = biomass + products

There are two conditions which determine conservation scenarios. They are individual masses for each element. They are C, H, O, N, P, K and it is equated by mass balance. In case of ionizing substances, the charges need to be balanced to achieve electric neutrality. The concept of described formalism is widely used in biotechnical formulations. It is the concept which describes the concentration of nutrients which are to be used to achieve required concentration of biomass or product in the fermenter. The growth yield is a primary parameter as far as the stoichiometry is concerned. This may be given as

$$Y = \frac{\delta_x}{\delta_s} \approx -\frac{\ddot{A}_x}{\ddot{A}_s}$$

Δx denotes the enhancement of the microbe biomass by using the substrate which is Δs.

Two substrate groups are needed for the growth of microorganisms. They are

- Catabolic substrates
- Conserved or anabolic substrates

The catabolic substrates are considered as the energy sources. The conserved or the anabolic substrates are considered as sources which provide biogenic elements which are required for formation of cell mass or materials. The catabolic substrates are of diverse nature and they include S^0 in case of sulfur oxidizing bacteria, NO^-_2 and NH^+_4 in case of nitrifying bacteria and also H_2 if a lithographic hydrogen bacterium is used. The consumption of these substances by the microorganisms is followed and accompanied by dissipation of chemicals as waste and they cease to exist as energy providing source. The anabolic substrates or the conserved substrates are incorporated into cellular components that means into biomass. These substrates can be reused. They are released into the medium as a result of cell lysis and they are readily available for the cells which survive. The anabolic substrates are basically non-carbon sources and they include trace elements, Iron, Magnesium, Potassium, Phosphor and Nitrogen. These substrates also include growth factors, amino acids and carbon dioxide in case of autotrophs.

The yield of biomass differs based on the selected microbe and the carbon substrate used. The concept of difference in the yield of biomass can be explained by mass energy balance theory. The heat released due to chemical oxidation or biological oxidation equals to the electrons gained and also the oxygen uptake. The production of heat due to the oxidation reaction gives 27 Kcal for available electrons equivalent. The number of available electrons for each carbon atom is known as carbon reduction degree or Y. It can be given as

$$CH_pO_nN_q + bO_2 = CO_2 + 0.5\ (p\text{-}3q)\ H_2O + qNH_3$$

$$Y = 4b = 4 + p - 2n - 3q$$

The available electron and the balance equation can be given as

$$Y_s + b\ (-4) = Y\gamma_x + Y_P\gamma_p$$

Y_s is the carbon reduction degree of the substrate

Y_x is the carbon reduction degree of the biomass

Y_p is the carbon reduction degree of the extracellular product

The available electron distribution among the oxygen i.e the electrons used for respiration, biomass and intracellular products. It is given by

$$\frac{4b}{\gamma_s} + \frac{Y_{\gamma_x}}{\gamma_s} + \frac{Y_{p\gamma_p}}{\gamma_s} = 1$$

A portion of the available electrons are sent to biomass from the substrate utilized. This is known as energetic growth yield.

$$n = Y_C \gamma_{x/\gamma_s}$$

The portion of internal energy of the substrate will be migrated into the product. This is known as energetic product yield.

$$\xi = Y_p \gamma_p / \gamma_s$$

The energy is required by the cells to grow and reproduce. The cells won't require growth to release or waste energy. The catabolic processes are a useless process and especially the process of respiration won't result in any growth.

The growth models can be classified into two types. They are deterministic and stochastic models. The deterministic model describes regular and determined processes. The stochastic process deals with random processes. The primary variables used in deterministic models are similar to the variable found in products, substrates, biomass concentrations and chemical kinetics. The stochastic model has the variables which include variance, frequency distributions, probabilities and many others. Stochastic model describes the probability of growth of bacteria under specific set of conditions. Every biotechnological process needs both stochastic and deterministic components. Many growth patterns exhibit deterministic model.

The biomass concentration and the number of cells are related and they can be given as

$$X = m X N$$

m is considered as constant and it shows the dry mass present in single cell.

N and X are considered as equal measures of growth.

The average mass and size of cells varies and they are dependent on environmental conditions and the organisms under study. The biomass is important to understand the mass flow controls, number of cells and some of the population studies in microbial population. The plasmid transfer and the mutation are considered as key factor which controls efficiency in any biotechnological process. The method to be used for determining the number of cells and biomass needs to be decided by taking into consideration several factors. The methods which gives us accurate measurements and which are automated needs to be chosen. The sophisticated methods are now in the market which has the capability of automated measuring of volatile substrates, methane, end products, alcohols, fatty acids, carbondioxide, oxygen, hydrogen and several other fermentation products which can be measured by, gas chromatography, liquid chromatography andmass spectrometry.

The variables at the primary state can directly be measured in culture. The variables obtained in the primary state can be recalculated as the characteristics of secondary growth. The characteristics of the secondary growth are gross growth rate, specific growth rate, degree of multiplication, biomass doubling time, growth yield, product yield, specific rate at which the substrate is consumed, specific rate at which the product is formed. All the characteristics measured are known as specific rates for each cell mass unit. These specific rates are less sensitive to the changes occurring in the biomass of the cell and hence considered as enzyme activity when studying kinetic derivations. The specific growth rate can be studied as the autocatalytic enzyme activity. The specific growth rate which is measured fr the dynamics ofbiomass has differences with the enhancement of cell number.

If the media contains ingredients with non-limiting concentrations, no inhibitors, sufficient physicochemical conditions, good mixing, then the biomass increase (dx) in an infinitesimally small time (dt) interval is directly proportional to the time interval and the biomass concentration (x) at the initial stages. This can be given as

$$dx = \mu x dt$$

μ is considered as the proportionality constant.

It was observed that exponential growth occurs in both eukaryotes and prokaryotes and not dependent on the morphological features. This condition holds good for fungi, protozoan's, plant and animal cells which were grown homogeneously. The difference in the exponential growth is only observed due to growth linked environmental variations. Precise measurement of the growth dynamics was given by Wang and Koch when studying *E. coli* which was cultivated in a computer linked cuvette equipped with double beam spectrophotometer. They observed that the lag in the growth is due to diauxie phenomena which nothing but the decrease of preferred oligopeptides. They also observed increase growth in succinate minimized mediums. The long term microbial cultures showed increased growth due to mutant auto selection and resulted in specific growth to the maximum in specific set of conditions.

Method	Principle	Pros	Cons
Plating	Colonies on petri plates	Low cost Higher sensitivity	Low precision More time consuming
Coulter counter	Counting and sizing is automated	Quick with automation, Size distribution pattern generated	Non cell solids can interfere
Image analysis	Computer assisted	Speed, size distribution generated, quantification by automation and manual	Higher costs
Microscopy-Manual	Hyphal length can be measured, cells can be counted	Biomorphological structures can be visualized, lower cost with more sensitivity	More time consuming, precision is low
Electrical measurements	Conductivity principle	Automation, quick and simple	Electrolyte interference, differences in conversion factor
Opacity	Light scattering principle	Quick, automated and simple	Interference due to wall growth, aggregation of cell and non-cell solids

Metabolic potential	Based on the rate of product formation or substrate added	Quick and higher specificity	Differences in conversion factor from biomass to measured rate
Fatty acids	Colorimetric reaction, Liquid chromatography or gas chromatography	Identification of mixed cultures, higher specificity	Differences in intracellular component and extracellular accumulation
ATP	Bioluminescence assay	High specificity and high sensitivity	Changes in intracellular ATP loop
DNA	Estimation of deoxyribose using colorimetric method	DNA in the cell is constant and higher specificity	Extracellular accumulation is possible
Folin Ciocalteau protein	Colorimetric reaction of tryptophan and tyrosine	Higher sensitivity	Possible extracellular accumulation, changes in biomass-protein ratio. Changes in cell protein
Biuret proteins	Colorimetric reaction- Peptides	High rate of uniformity	Changes in biomass and protein ratio
Particulate organic carbon	Measurement of carbondioxide post separation of cell	Higher rate of precision and sensitivity	Interference from non cell solids and dead cells
Wet biovolume	Linear dimension of colony or cells which are pelleted	Quick and simple	Changes in biomass density (Wet)

Wet mass	Mass of material which is separated	Quick and simple	Changes in biomass
Dry mass	Mass of dried and separated solids	Direct estimate	Interference from noncell solids and dead cells

Model proposed by Monod

This model is simple and it played an important role in understanding the microbial kinetics. This model considers conservation of mass condition, substrate, biomass formation, yield factor which is constant, specific growth rate. The Monod's model is an unstructured model and it gives only a minute manifestation of the phenomena involved in growth. This model is used to explain steady state growth and balanced growth. If we want to analyze the non-steady grow and unbalanced growth, then it requires strict formulation of models which are structured.

Balanced growth can be seen as a proportional growth in every component of the cell and no variations are seen. The non-steady and steady state concepts are derived from the enzyme and chemical kinetics. The steady state refers to the constant reaction rate between break down and formation of products as in case of enzyme substrate reactions. Generally, when studying microorganisms, we are studying steady state and the rates in totality remains constant. In case of open system cultures, the total and specific rates remain constant. In a batch culture of the closed system, these rates are constant only during exponential phase. The non-steady state in the process of biotechnology and microbiology is always clear. They allow different hypotheses which are to be tested and they yield adequate data about the objects under study. They contain higher value in biotechnology. They are capable of providing tools for regulation of bioreactor. It displays higher productivity and efficiency. In case of long term processes, non-steady state becomes unbalanced as there are changes in composition of the cell. In case of heterogenous population, this study becomes complicated. The cells from the first stage are different from the cells obtained from the second phase. This is called as transient steady state. The macromolecular cells are classified into two groups. They are P components and U components. The P components are the primary cell constituents and they are required for intensive growth of cells and the biomass. The U components are required for the survival of the cell. The U components are responsible for cell survival even in presence of inhibitors. Some of the examples of P components are ribosomal proteins, rRNA and enzymes which are required for metabolic pathways. On acceleration of growth, the contribution of U compounds on biomass diminishes. Some of the examples are reserved substrates, high affinity transport systems, protective pigments and enzymes.

The P and U components are to be expressed in the terms of total cell mass. If one P component is increased then there would be a drastic increase in more P compounds too. The ratio between P and U components is regulated by the substrate which acts as limiting factor in a given environment. The cell cycle consists of certain growth conditions which are said to be in steady state and it is classified into three phases. They are

- Growth of cell without taking into consideration the replication of chromosome from initial stage to the critical stage.
- The replication of DNA called as C phase
- The daughter cell separation called as D phase

The C and D periods are constant and these periods take around 20 to 40 minutes and are not dependent on environmental conditions. The first phase is based on the culture conditions. The cell components are classified into three categories according to Cooper. They are

- The exponential accumulation of small molecules, ribosomes, RNA and proteins.
- The replication of DNA in the cell is in linear fashion.
- The surface of the cell is composed of peptidoglycans which are synthesized during cell cycle.

The microbial growth is based on three important features. They are

- The supply of substrate. It can be single term addition or it can be supplied continuously.

- The microorganisms which are grown will be eliminated. In some cases, the growth would be significant and in some cases, the growth may be absent.
- Spatial gradients and their magnitudes. It can be of heterogenous system or it can be homogenous system.

The microbial growth kinetics is based on the fermentation process design. The fermentation design is based on the cellular processes. There are two key parameters involved in this design process. They are productivity and yield. In describing productivity, the time constrain is taken into consideration and the production formation rate. This is expressed in grams/ liter per hour. The yield explains the amount of product obtained by consuming substrate in case of expensive raw materials. In order to understand the kinetics, there are certain procedures to be followed in modeling. They are

- The stoichiometry of the reactions and the number of reactions are to be considered. This is called as model complexity.
- The cell reaction rates along with variable functions needs to be specified. This is called as setting up of kinetics.
- The kinetics needs to be combined with the reactor model. The inflows and outflows of the bioreactor, the change of concentrations in metabolic products, biomass and substrates needs to be specified in any model. The mass balances of the reactor should be specified in this model.
- If the experimental data does not fit with the simulation, then the complexity off the model needs to be re-framed.

The rate and the yield coefficients are important factors in understanding the cell growth kinetics. The inflow of the substrate into the culture containing cells can be detected by changes in culture concentrations. The outflow of the products is detected by their relative concentrations.

REFERENCES

1. Sikyta, B., Prokop, A., Novfik, M. (1974) Advances in Microbial Engineering, Part 2, Biotechnol. Bioeng. Symp. 4, New York.
2. Umbreit, W. W. (1974) Adv. Appl. Microbiol. 17, 1.
3. Volesky, B., Votruba, J. (1992) Modelling and Optimization of Fermentation Processes. Amsterdam.
4. Vossius, V. (1979) Biotechnol. Lett. 1, 187.
5. Votruba, J., Sobotka, M. (1992) Folia Microbiol. 37, 331.
6. Wegner, H. C. (1979) Biotechnol. Lett. 1, 145.
7. Bergter, F. (1978) Z. Allgem. Mikrobiol. 18, 143.
8. Bu'Lock, J. D. (1965) The Biosynthesis of Natural Products. An Introduction to Secondary Metabolism.
9. London.
10. Bu'Lock, J. D., Hamilton, D., Hulme, M. A., Powell, A. J., Smalley, H. M., Shepherd, D., Smith,
11. G. N. (1975) Can. J. Microbiol. 11,765.
12. Caldwell, J. Y., Trinci, A. P. (1973) Arch. Microbiol. 88, 1.
13. Contois, D. E. (1959) J. Gen. Microbiol. 21, 40.
14. Deindoerfer, F. H. (1960) Adv. Appl. Microbiol. 1, 321.
15. Dosko~il, J., Sikyta, B., Kagparovfi, J., Dosko~ilov~, D., Zaji~ek, J. (1958) J. Gen. Microbiol. 18,302.
16. Gaden, E. L. (1959) J. Biochem. Microbiol. Technol. Eng. 1,413.
17. Hattori, T., (1985) Rep. Inst. Agr. Res. Tohoku Univ. 34, 1.
18. Herbert, D. (1961) The chemical composition of micro-organisms as a function of their environment. In: Symp. Soc. Gen. Microbiol. 1 l, Cambridge.
19. Kretschmer, S. (1978) Z. Allgem. Mikrobiol. 18, 701.
20. Lineweaver, H., Burk, D. (1934) J. Am. Chem. Soc. 56, 658.
21. Luedeking, R., Piret, E. L. (1959) J. Biochem. Microbiol. Technol. Eng. 1, 431.
22. Mfilek, I. (1976) Physiological state of continuously grown microbial cultures. In: Continuous

23. Culture 6: Applications and New Fields, Chichester.
24. Maxon, W. D. (1955) Appl. Microbiol. 3, 110.
25. Monod, J. (1949) Ann. Rev. Microbiol. 3, 371.
26. Moser, H. (1958) The Dynamics of Bacterial Populations Maintained in the Chemostat, Washington, D. C.
27. Pirt, S. J. (1966) Proc. Royal Soc. B, 166, 369.
28. Pitt, S. J. (1967) J. Gen. Microbiol. 47, 181.
29. Pirt, S. J. (1975) Principles of Microbe and Cell Cultivation, Oxford.
30. Righelato, R. C. (1975) Growth kinetics of mycelial fungi. In: Filamentous Fungi 1, London.
31. Schuhmann, E., Bergter, F. (1976) Z. Allgem. Mikrobiol. 16, 201.
32. Teissier, G. (1942) Reo. Sci. 80, 209.
33. Balbas, P., Soberon, X., Merino, E., Zurita, M., Lomeli, H., Valle, F., Flores, N., and Bolivar, F. (1986). Plasmid vector pBR322 and its special-purpose derivatives – a review. Gene, 50, 3–40.
34. Casali, N., and Preston, A. (ed.) (2003). *E. coli* Plasmid Vectors: Methods and Applications (Vol. 235, Methods in Molecular Biology), Clifton NJ: Humana Press.
35. Shizuya, H., Birren, B., Kim, U.-J., Mancino, V., Slepak, T., Tachiri, Y., and Simon, M. (1992). Cloning and stable maintenance of 300-kilobase-pair fragments of human DNA in Escherichia coli using an F-factor-based vector. Proceedings of the National Academy of Sciences U.S.A., 89, 8794–8797.
36. Kehoe, J.W., and Kay, B. K. (2005). Filamentous phage display in the new millennium. Chemical Reviews, 105, 4056–4072.
37. Lipovsek, D., and Pl ¨uckthun, A. (2004). In-vitro protein evolution by ribosome display and mRNA display. Journal of Immunological Methods, 290, 51–67.

CHAPTER 3

CULTURE SYSTEMS

INTRODUCTION

The microbial cells can be cultured for a specific time under steady state conditions. This culture is often referred to as continuous cultures. The cells remain in a constant physiological state and the end product will be uniform. The population density can be changed as per the user requirement. The important feature of the continuous culture is the steady state.

Classification

The continuous culture systems are differentiated based on the operation principles. In case of open systems, the microbial cells are subjected to continuous removed with effluent fluid intact and it is based on the rate of input of new cells. In these conditions, it is easy to attain a steady state condition. If we take closed system, the cells are subjected to retention and this result in increase of cells at a steady rate. The main characteristic of the open and closed system is that the open systems can be operated in a state defined as 'dynamic steady state' and the closed systems cannot exhibit dynamic steady state.

Open continuous culture systems

In these systems, the composition of the culture is kept constant. The commonly used continuous culture systems can be called as homogenous continuous culture systems. These are classified into single state systems and multi-stage systems. In multi-stage systems, two or more vessels are installed in a chain like fashion. The multi-stage system may comprise of a single chain along with medium inflow which enters into the first stage. This system may also form complex chain which comprises of several medium inflow inputs or feeds. The composition of the medium in vessels and the volumes in the vessels can be similar or in some cases different. The advanced multistage system is the one having complex chain with mutual culture exchange facility. The open systems which are homogenous may be modified with feedback of cells and it might be partial.

The continuous culture systems which are single state can be classified into two types. This classification is based on the procedure used to attain or get the steady state condition. They are the chemostat as proposed by Novick and Szilard (1950) or bactogene as proposed by Monod (1950) and the turbidostat as proposed by Bryson and Szybalski (1952). In the chemostat, the limiting substrate is kept constant and this substrate is crucial for maintaining cell concentration at a constant rate in the culture medium. The flow rate is controlled and it is designated as F along with volume which is constant and it is designated as V. The dilution rate D is the ratio of F and V. In normal steady state conditions, D is proportional to the population growth rate which is designated as μ. The dilution rate is kept below the μ_{max} which is the maximum population growth rate.

In turbidostat, the cell population in the medium is maintained via rigorous monitoring with the help of turbidity or optical density of the culture whilst adjusting flow rate. When the population growth increases, the turbidity or the optical density increases. Once the threshold value is attained the fresh medium is allowed to enter. If the turbidity or the optical density is decreased, the entry of medium is stopped. Therefore, it is easy to maintain culture with D values almost equal to μ_{max}. Turbidostats have two disadvantages. The first disadvantage is the physical limitation and the second is that the obtainment of low steady state. These two disadvantages can be managed by installing two feed streams.

Classification of continuous culture systems

A Homogeneous- open continuous systems-single stage

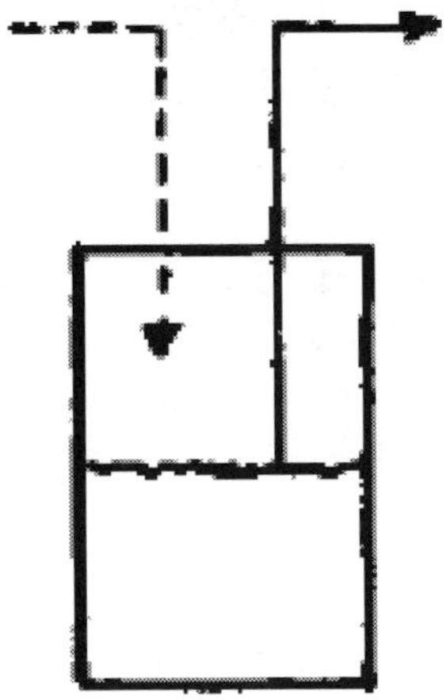

Simple (Turbidostat, chemostat)

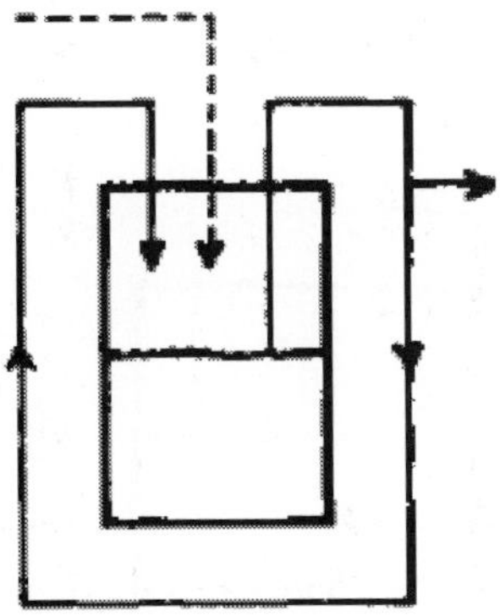

Simple (partial feedback)

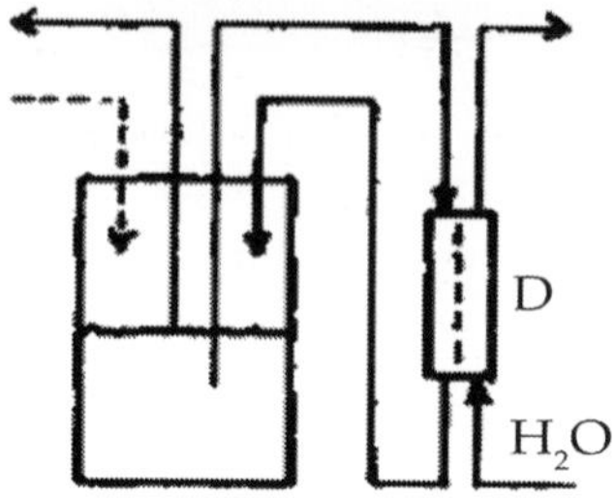

D-dialyzer

Multistage

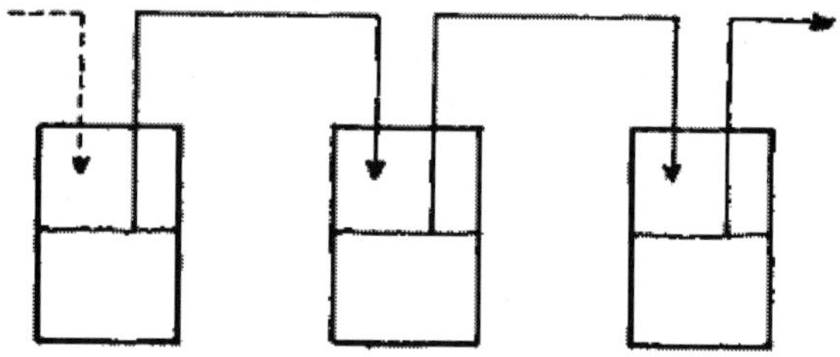

Simple chain (inflow into first stage)

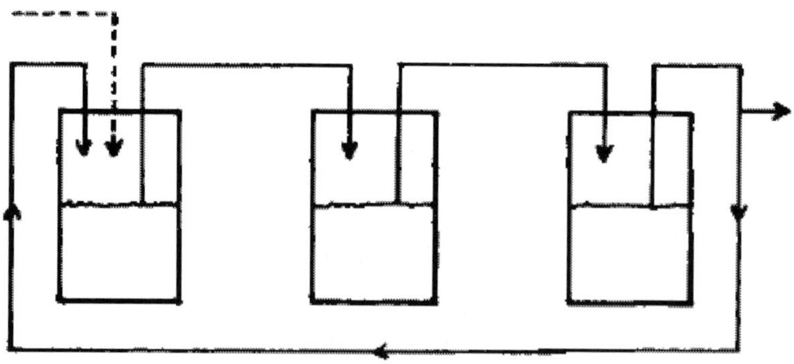

Simple chain (Partial feedback)

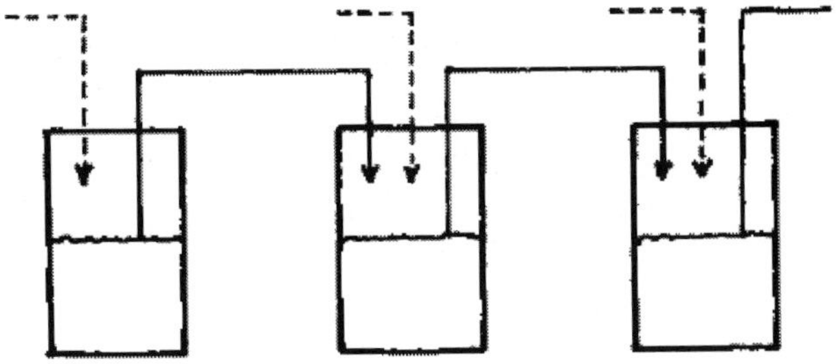

Complex chain (inflow into more stages)

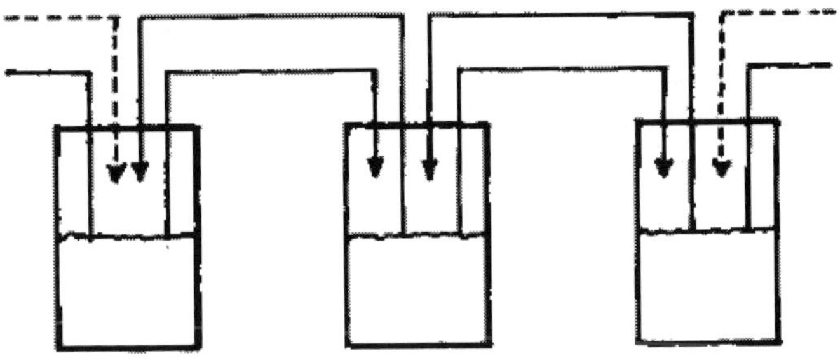

Complex chain (inflow into more stages-mutual exchange of culture)

B Heterogeneous

Simple tubular (solid or fluid substrates)

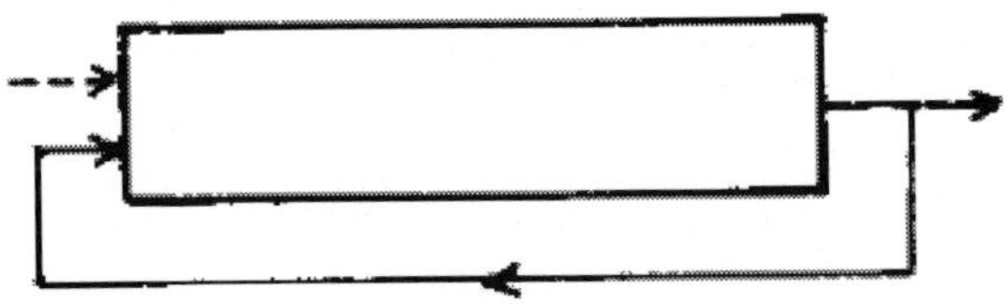

Tubular (Partial feedback)

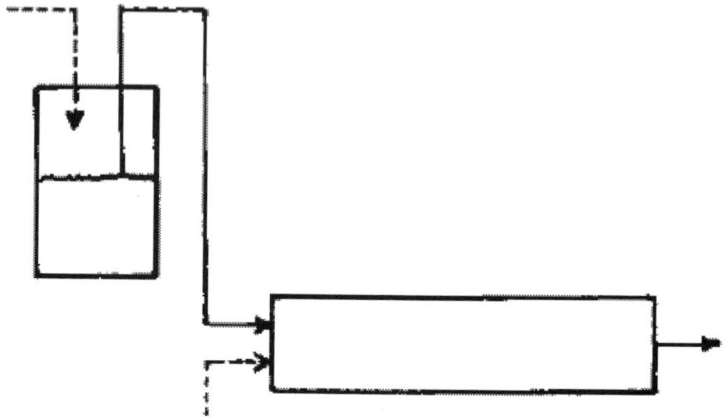

Tubular (inoculation from chemostat continuously)

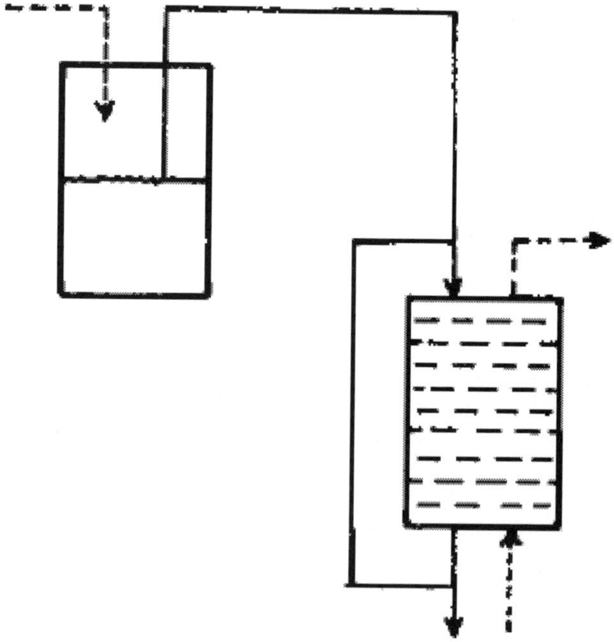

Counter-current system

Closed continuous systems

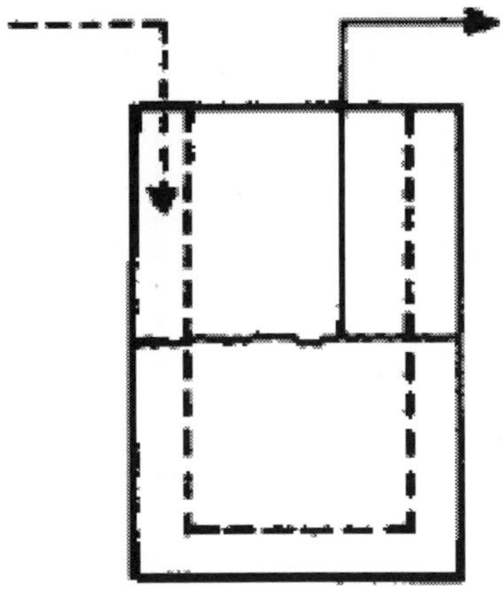

Vessel equipped with permeable walls

Microbial growth on walls (Medium will flow above the culture)

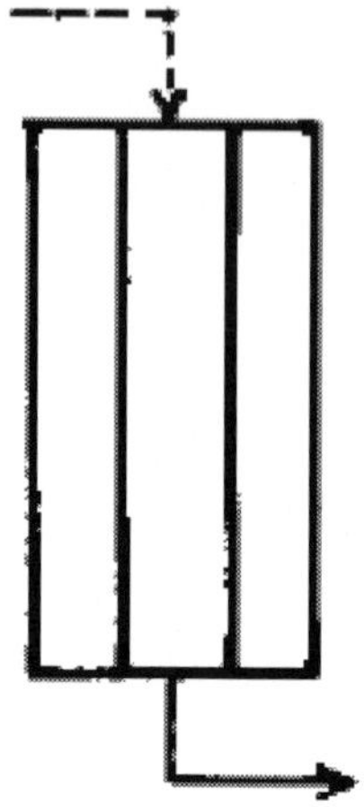

The medium flows interior

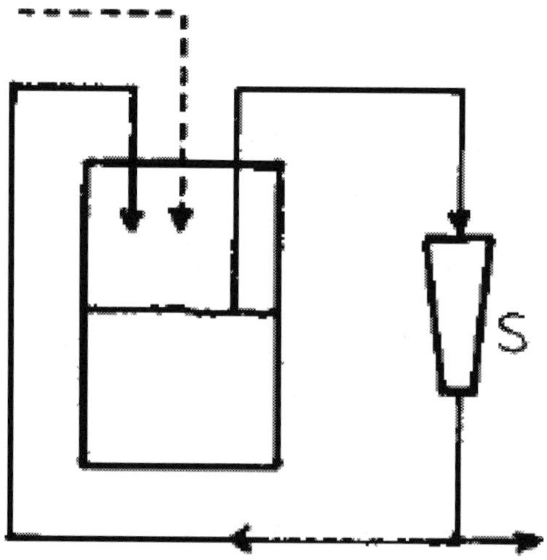

S-separator

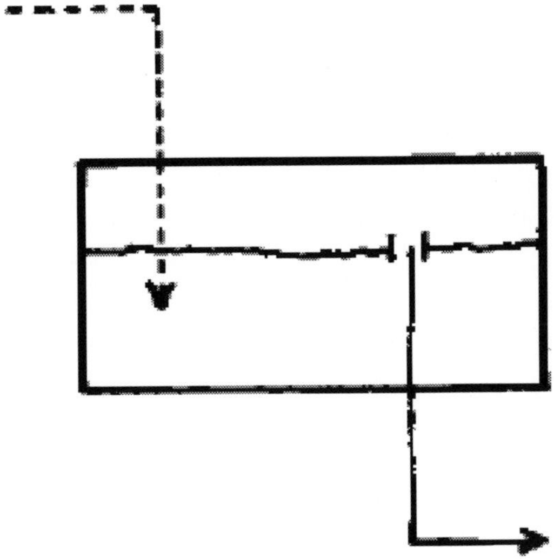

Growth of cells is on the interface (gas-fluid)

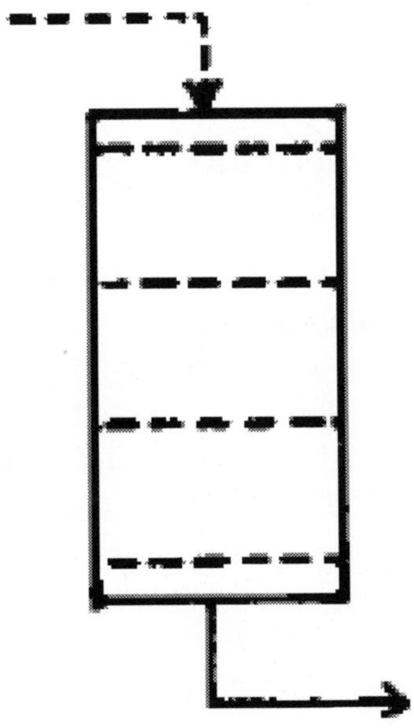

Fluid solid

The single stage system is used to culture cells with well defined medium in order to obtain higher amounts of biomass or the en products connected to the microbial growth. It is utilized in genetic, physiological, and biochemical research and cell mass production. This system is used to get a simple product which is formed as a result of a single metabolic transformation. This system can be used for improving the medium or for optimization of the medium. In this system, one parameter of the culture can be changed while keeping the other parameters constant. The composition of the medium can be systematically developed whilst observing variations in the steady state. The best strategy is to use continuous flow enrichment methods for selection of microorganisms. The enrichment of chemostat helps in selection based on various specific growth rates and concentrations of substrate. The continuous flow enrichment method is a proven technique in case of enzyme and single cell protein production. The production in the continuous culture is regulated by the lower microbial concentration in the bioreactor. The dilution rate needs to be lower than the maximum specific growth rate. This problem can be solved by removing a portion of cells from the fluid and feeding back to the bioreactor. The cells are recycled in the continuous culture and they are retained, so there will be an increase in the concentration of biomass in the bioreactor.

Recycling of cell by means of settling does require flocculent strain. The dilution rate is regulated by settling velocity of the strain. The centrifugation technique is not advisable in this case, since it is costlier. The membrane filter systems are best in enhancing the cell mass. The broth takes tangential flow along or across the membrane filter and the deposition of the cell on the surface of the membrane is reduced, hence high productivity rate is achieved. The membrane filters are fitted with the spin rotor and in some cases attached to the cartridges. The membrane system when attached to bioreactor produces culture of higher density and it removes metabolic products and cells get recirculated.

Multi-stage continuous systems

In this system, the individual systems are connected maintaining heterogeneity if we take as a whole but each cell exhibit homogeneity. The multi-stage system can form a simple chain. The vessels can be joined in parallel or in series or in the form of cascades. Each vessel feeds the other in only one direction. Vessels are designed with different volumes if varied rates of flow are required. The character of multistage is achieved by dividing single vessel both horizontally and vertically and into several floors and chambers. The fresh medium is supplied into first, second and third stages and so on as per the requirements especially when complex chains are used with an option of multiple substrate addition. The medium which is to be fed have varied compositions. The selected material which is to be fed need not necessarily be a liquid. If the dilution is to be prevented, then the powder form of media should be used. The part of the suspension can be fed into the same vessel or the vessel in the preceding stage. The remaining cell along with the substrate left is allowed to flow for processing. The system condition is decided by the ratio between suspension and the concentration of the cells. It also depends on the concentration of the cells and the feedback ratio. The concentration of the cells is more than the out flowing medium concentration. This concentration is decided by the degree of feedback. This process will help in good flow rate and increased production of cells.

The two-stage continuous culture provides advantages of both single stage culture and batch cultures. Every stage exhibits homogenous culture. The first stage can be conditioned for the growth of cells and the second stage is meant for formation of product. The multi-stage continuous culture can be used in the plant cell culture. This method is proved with Nicotiana tabacum. The cells which are observed in the second stage showed enhanced level of phenolics than the cells observed in the first stage. In this case, when glucose is fed especially in the second stage, there is an appreciable increase in the cell division and the phenolics production is diminished. In the same way when phenylalanine is added, there was an increase in the biomass and the phenolics production is reduced drastically. It is proved through these experiments that the multistage continuous cultures can be employed to increase the production of secondary metabolites and also in framing certain mechanisms for controlling the formation of secondary metabolites. The system which comprises multiple substrate addition in case of two stage system can provide several possibilities both for practical and experimental use.

This process was commercially tested in Sweden for production of microbial biomass while using potato wastes. *Saccharomyces fibuligera* is used to produce amylase in the first stage and the effluent is allowed to flow into the second stage. The second stage is supplemented with *Candida utilis* and *S. fibuligera.* To produce microbial fat, *Rhodotorula glutinis* was used and the two-stage system was installed. In this process, the first stage was designed to produce higher biomass growth. This system is crucial in disposing waste materials, utilized and transformed substances. The fluidized bed can retain higher biomass concentrations on a solid. The two-stage system was installed for alphs amylase production using *Bacillus amyloliquefaciens.* The activity of amylase and the productivity in terms of volume were tested with two stage system, single stage and batch systems. The chemostat in the single state has given higher enzyme productivity. In this case the continuous culture systems were weak as compared with batch systems.

The continuous systems which are multi-stage can work with two directions in a given point of time. This property allows the opposing gradients which solutes posses at end of chain. This is called as gradostat. The solutes and cells are pumped via tubes in only one direction and in some cases through gravity. The volumes in the vessels are same and the rate of flow in any direction is same. The residence times vary as per the chain position and it is bidirectional. The vessels which are located at the centre contain long residence time than the vessels at the ends. The non-limiting conditions do exist for microorganisms and hence the growth is greater.

Heterogeneous continuous open flow systems

These systems are characterized by the non-mixing of the fluids which occurs in the flow path. It can be called as imperfectly mixed system. This is a multi-stage process with several number of vessels and there will be transition between them continuously.

Tubular systems

The vessel exists in various forms and they can be oriented vertically or horizontally. The coils are equipped with degasifiers and they are used for the anaerobic process to take place. The horizontal tube contains rotating discs and they are closely arranged to a shaft. This type of systems is used in waste water treatment. The biological slime is allowed to grow on the disc and also submerged so as to facilitate nutrient absorption. It consists of compartments which are cylindrical and it is also sealed. Overflow hole is arranged on the upper half. It is useful in mold cultivation. Tubular bioreactor with a thin layer is used for the culture of low foaming microorganisms. Some modifications can be made for the design. Some of the variants are thin-layer bioreactor, thin layer film and agitated tubular bioreactor. In waste water treatment, the commonly used bioreactor is the pneumatically aerated tubular reactor. Air is allowed to pass over the bioreactor length. Rotary internal coils are equipped along with internal discs. Constant injection of inoculum is needed to achieve a steady state. In order to maintain continuous feeding of inoculants, chemostat is used.

Tower systems

The continuous tower fermentation technique was developed in Britain in 1960. The application of this technique is limited. Yeast generally aggregate in the form of flocs. This property makes settling easier. The settling zone is incorporated inside the bioreactor and it is free of turbulence which is due to carbon dioxide. The accumulation of yeast is simple easy and hence the conversion capability will be increased drastically. There is no need of cell recycle and sedimentation. The additional machinery can result in microbial contamination. In this system, the bioreactor's space is divided into several floors. The cells with the medium are allowed to flow in a vertical fashion inside the bioreactor. The liquid or the gaseous phase which is lighter generally takes an opposite direction. This system is more effective and some portions of the culture are fed back. The choice of using vertical or horizontal bioreactor is based on the type of the microbial process to be used. It is observed that the ethanol fermentation using yeast cells is higher in case of horizontal flow channel bioreactor than its vertical counter parts. The higher rate of fermentation in case of horizontal bioreactor is due to yeast cell accumulation by sedimentation and partial mixing in liquid flow. This system can also be used for solid substrate fermentation especially when cellulosic material is used. The arrangement of bioreactor can be in horizontal or vertical position. In case of vertical configuration, the chambers are arranged one above the other. This pattern forms a tower like system. In case of horizontal type, a tubular bioreactor is established. In the first stage, the substrate and solutes are fed separately followed by rigorous mixing. The higher rate of microbial growth takes place. The rate of addition of the solutes and the solid substrates depends on the cellulose decomposition rate. The fermented solutes though partly fermented are allowed to pass into the later stages. In case of tower system, the materials which are present in the chambers are allowed to aerate with the help of upward flow of air and this process is assisted by agitation mechanically. In case of horizontal tubular reactor, the material which is fermented is passed via a tube with the help of running belt or spiral belt. This reactor is loosed mechanically and also aerated accordingly. The retention time is decided by the rate of decomposition of cellulose.

Closed continuous culture systems

In these systems, microorganisms remain in the system as opposed to the open systems. There will be an increase in the dead cells and hence one or more limiting factors do emerge. The culture cannot achieve dynamic steady state. The entire process is a prolonged process in batch state and the change in the biophase with respect to time is pronounced. These systems are not allowed as true continuous flow processes.

Total feedback

The total feedback of cells is replaced by the partial feedback of cells and this is the characteristic feature of closed systems. The fresh medium is fed into the vessel and the cell suspension is also added to the medium. The medium and cell suspension continuously circulates in the vessel. The cells are harvested from the used fluid and the medium is feedback to the system till all the nutrients are used up. These systems are useful in growing *Streptococcus lactis*. The cultures along with cell recycle are achieved by a fiber unit which is connected to the reactor. The target is to get higher rate of cell densities resulting in superoxide dismutase. This growth of cells is devoid of the inhibitor mainly lactic acid.

The inhibition of the product is the key factor which limits microbial processes. If the productivity is to be increased, then it is mandatory to remove the product produced by the cells. Techniques such as solvent extraction and vacuum evaporation are used to remove these products. The substrate is added to continuous bioreactor and the water is added later.

When compared to the old non-dialysis processes, this process results in high concentrated substrate. Through this process, minimal substrate is utilized to form product and the substrate is subjected to sterilization via membrane passage. The bioreactor is installed with microfilter which is a rotating cylinder and it allowed the removal of cells. The continuous extraction of the inhibitors within the filter bioreactor produced greater density of Streptococcus faecalis than the batch mode of cultivation. A roto-fermentor was employed to produce higher amount of ethanol from the substrate glucose. The concentration polarization and fouling problems are solved by installing rotating membrane module.

The cells are retained physically in the immobilized cell bioreactors. The process of physical retention is achieved by gel encapsulation, chemical coupling, and natural attachment. The immobilized bioreactors are useful because of some advantages such as reduction in substrate and end product inhibition, reduced product recovery costs, waste treatment, operating and capital costs, high rate of volumetric productivity and continuous operation on long term. Active immobilization or the passive immobilization of cells limits the loss of cells from the bioreactor. In case of active immobilization, the cells are mixed with chemical agent and are fixed by physical or chemical means. In case of passive immobilization, the flocs or films of cells which are surrounded by support material are utilized. The inoculation is done in the normal way and the inert particles are added.

The continuous processes which employ the immobilized biocatalysts can be done by using different types of bioreactors. They are classified into two main types namely bioreactors equipped with piston flow and mixed bioreactors. The selection of bioreactor is based on several factors. They are products or inhibitory products, substrate concentration, oxygen supply, solution adjustment requirements, microbial process type and the type of biocatalyst selected.

Packed bed bioreactors

The bioreactor equipped with a piston flow is nothing but a column and the backflow mixing is not allowed and the entire mixture is made homogeneous thermally. The substrate can allow flowing either from top to bottom or from bottom to top. The biocatalyst which is present in the column is compressed occasionally and so the layer is divided by the help of baffles while maintaining equal pressure. It is beneficial to arrange the columns inside the battery with one column differentiated as main reactor and the others are differentiated as safety vessels. Once the main column is used up, the safety column is used as main column and the other new safety column is automatically switched on. The upward flow reduces immobilized biocatalysts compression and this creates the distance among the biocatalyst particles.

The organic pollutants which are present in waste gases are eliminated by using a biological filter bed technique. The bed consists of a specific material which has to support the microbial growth preferably on a solid surface and also the bed supplies the microorganisms with nutrients which are inorganic in nature. Carbon dioxide and water are the produced due to oxidation of the organic compounds. The solid phase and the microorganisms which are present are arranged in such a way that the process of ageing is prevented. This helps in maintaining high rate of activity over long periods. The apparatus constitutes of columns made up of glass and they were separated into five stages. Each stage is supported by a perforated sieve plate arranged at the bottom. This package is generally inoculated with a sludge suspension obtained as an effluent from municipal waste water treatment plant. The humidity of the filter bed is maintained by arranging a tap water which has the capability of spraying from the top for each stage. The process constitutes of the unit processes, anaerobic filter which is bell-saddle-packed and activated sludge nitrification systems. In the first stage, the anaerobic filter is used to remove the organic material without liberal of methane gas. Good amount of organic matter reduction is possible in activated carbon anaerobic filter which is also granular. The success provided by activated carbon anaerobic filter is attributed to its power of activated carbon which helps in sequestration of the wastewater components which are toxic to the anaerobic microorganisms especially when mixed cultures are used. The activated sludge nitrification system helps in complete oxidation of ammonia and good in polishing of the effluent.

Fluidized bed bioreactors

They are generally conical in shape and also look like it is tapered to the bottom. The upper part which is broadened helps in maintaining linear and decreased up flow velocity and pressure gradients. The particles settle in the narrow part which is at the lower ends. These are best used for active and passive immobilized cells. In passive immobilized cells, the cells along with the solid substrate are generally dense as compared to the aqueous media and they can be allowed to retain with the column so the nutrients are fed in the ascending fashion. The particles transform into fluids and rapidly move. The rapid motion generally carries two effects. The solute exchange is improved between cells and medium where the cells are attached to the substrate. The cells which are formed on the film surface can be removed by the process of attrition. The biological fluidized bed bioreactors are effective in treating wastewaters loaded with organic substances. These bioreactors provide ample surface for the colonization of microorganisms and this gives a chance to obtain higher rates of biomass concentration. These bioreactors can yield higher substratc consumption rate. It performs well in low temperatures and allows shorter retention times even in low strength wastewaters.

Fluidized bed bioreactor is more beneficial than the fixed bed bioreactor in case of ethanol production. This benefit is due to the enhanced rate of transportation of carbon dioxide gas.

Film bioreactors

The biofilm systems help in retaining slow growing microorganisms in a flow reactor which is continuous. The retention time especially in the liquid phase can be chosen freely without considering the specific growth rates. The high microbial concentration can be achieved due to diffusional limitations. The conversion rates of the substrate are proportional to the film surface area and are independent of microbial biomass.

The fixed film bioreactors which are anaerobic down flow stationary can retain the microorganisms in the bioreactor on supports and are used to digest wastes which are present in suspended solids which are loaded with organic material. These systems are capable of producing higher rates of methane. The feed will be pumped from the top and the resulting effluent can be taken out from the bottom for either recirculation or disposal. One of the other important film bioreactors is the rectangular film bioreactors. Porous stainless-steel fiber sheets or cloth are used for the entrapment of microorganisms. These are arranged vertically and in parallel patterns in case of horizontal bioreactor. Multi-stage or the single stage system can be used for production of various metabolites.

Hollow fibre bioreactors

The geometry which is also called as immobilization geometry comes with a cylindrical configuration of ultra-filtration membranes which are asymmetric and hollow. These membranes contain anisotropic structures and their inner layer is ultramicroporous. The layer is 0.5 micrometer thick and it is surrounded by polymer matrix and it is open celled and macrophorous. The inner wall is dense and it is a semi permeable membrane with pore diameters in nanometers. The matrix exhibits large hydraulic permeability and contain 90% of void space.

A fibre is generally epoxied inside the glass shell and the inoculation is done in the bioreactor with the help of access ports present in the outer shell. The nutrients which are in the liquid phase flow via fibre lumen and gets diffused along the ultra microporous membrane to the cells in the membrane region. The metabolic products from the cell are excreted within the fibre wall especially in the extracellular space. The products get diffused into ultra microporous membrane and then they gets into fibre lumen. The products are sent to fibre lumen and then there establishes a continuous process of transforming substrates to products. The process of continuous flushing can be exploited to supply the oxygen and also removal of carbon dioxide. Dual hollow fibre based bioreactor systems were used for immobilizing microorganisms. The immobilized microorganisms grow aerobically in these kinds of systems and they grow in higher densities.

Overproduction and regulation

Some of the mechanisms which regulate the over production of microorganisms are mostly enzyme based. The regulatory mechanisms are substrate induction, catabolite regulation via repression and inhibition, feedback regulation via repression, inhibition and pathway modifications, amino acid regulations, energy charge regulation and permeability control.

Substrate induction

Enzymes which are released from microorganisms only when the substrate is available in the provided medium. These enzymes are termed as inducible enzymes. The inducer is an analogue of the substrate. If the inducer is available in the medium, various inducible enzymes are produced by the microbe. This process occurs only when the metabolism pathway is dependent on sequential induction. In this case the microorganisms produce enzymes if substrate is present. Intermediates are produced while the enzyme acts on substrate. This process is continued until the substrate and the metabolism is exhausted. There are other set of enzymes which are produced even in absence of substrate. They are called as constitutive enzymes. The enzyme induction allows the organism to react rapidly and sometimes even in seconds if the substrate is present and so the enzymes which are not needed are not produced.

Catabolite regulation

The carbon compounds which are not inducers have profound effect on the synthesis of proteins. If more than one carbon source is available, the microorganisms select the carbon source which allows the organism to grow rapidly. The enzymes needed for non-preferable carbon sources are repressed and will not be produced. This effect was observed when the microorganisms are supplied with both glucose and lactose. This effect is rightfully called as 'glucose effect'. As the glucose is preferred carbon source to the organism, then it produces enzymes accordingly. It suppresses lactose utilization until the glucose molecules are exhausted. After several years, it was found that a catabolite is responsible for the glucose effect and hence the name was changed to "Catabolite repression". One of the active catabolite which is involved in the repression is cyclic adenosine monophosphate (cAMP). In many cases minute amount of cAMP gets accumulated in order to promote rapid growth of the microorganisms. The catabolite repression is proved to be reversed by cAMP.

Feedback regulation

Feedback regulation is also called as the end product regulation. Feedback regulations are very crucial in regulating biosynthetic and anabolic enzymes but the catabolic enzymes are controlled by catabolic or induction regulators. There are two types of feedback regulation. They are feedback repression and feedback inhibition. These two products are involved in adjusting the production rate of the end products.

Feedback inhibition

Any metabolic pathway results in an end product and this will inhibit the enzyme action. The substrate and the inhibitor may not resemble each other so this relation is called as allosteric but if there is resemblance it is called as isosteic inhibition. In case of isosteic inhibition, the substrate and inhibitor share the same molecular conformation. Feedback inhibition may be explained at enzyme level through the enzyme molecular structure. Each enzyme will have two protein sub units. One binding site present on the subunit have affinity towards substrate and the other binding site present in the second subunit have affinity towards feedback inhibitor. If the inhibitor bins to the enzyme, the structural conformation of the enzyme changes and it no longer can bind to the substrate. This is called as allosteric effect.

Feedback repression

The feedback repression reduces the rate of enzyme synthesis. This is crucial in avoiding vitamin over production which is generally needed in lower amounts. The feedback repression can be achieved with the derivative of the end product. This repression is slow in removal and also introduction.

Regulation in branched pathway

The branched pathway can bring about two end products and there would be a problem if one end product inhibits the synthesis of other. Several patterns concerning feedback inhibition have been studied in cases of branched pathways. They are concerted or multivalent feedback regulation, cooperative feedback regulation, and cumulative feedback regulation, compensatory antagonism of feedback regulation, sequential feedback regulation and isoenzymes with certain regulatory effectors. In concerted or multivalent feedback regulation, the two end products will have no negative effect. If both the end products unite, they act as inhibitors. In case of salmonella, the branched sequence leads to production of pantothenic acid, isoleucine, leucine and valine. In cooperative feedback regulation, the end products if acts individually can induce a weak inhibiting power, but if both the end product acts synergistically, then there would be total inhibition. In case of cumulative feedback regulation, one end product has the capacity to totally inhibit the enzyme without the use of second end product.

Amino acid regulation of RNA synthesis

The synthesis of proteins and the RNA ceases if the mutant is not given the required amino acid which is supplied in the medium. In this case, the cells stop to overproduce RNA which is unwanted. These economic strains are termed as stringent. These mutant strains are allowed to produce RNA even in the absence of amino acid.

Energy charge regulation

The cells have the capability to regulate the energy which is required for any reaction to takes place. The energy compounds such as adenosine monophosphate (AMP) and adenosine triphosphates (ATP) are generally produced in the process of catabolism. The energy charge can be calculated by computing the ratio between the ATP and half of adenosine diphosphate (ADP) to the ATP, ADP and AMP. The charge which cell possess ranges from 0 to 0.1.

Permeability control

Permeability control allows the microorganisms to hold the essential molecules inside the cell and it permits some selective molecules from the environment. This process occurs at the cell membrane. Solute molecule can pass through a lip protein complex and there is a need of driving force. The solutes are transported through active transport which is mediated by a carrier and passive transport via diffusion.

Passive transport

The electric charge and the concentration gradient are the driving forces for the passive transport. In case of non-electrolytes, the concentration gradient plays a major role. If ions are used, then the electric charge which gets established inside and outside of the cell membrane is crucial.

Transportation mediated by carriers

There are three types of mechanisms namely facilitated diffusion, active transport and group translocation. In case of facilitated diffusion, the concentration difference of the solute is the main driving force. The carries enhance the passage rate across the membrane. In case of active transport, the material required by the cell is accumulated inside the cell against concentration gradient. The energy which is required for solute transport is supplied by the enzymes named permeases. In this case, the solute is not altered. The permeases act on only some compounds and in many instances, they are controlled. In group translocation, the solute when getting transported is chemically altered and then it gets accumulated inside the cell. The carrier molecules are involved in catalyzing the group transfer reactions with solute substrate. The group translocation constitutes two activities. They are entry and exit process. The carrier mediated transportation is crucial as it is selective and it is also considered as the rate limiting step in carbon metabolism.

The metabolic control of metabolites which are primary can be deranged and this is done with some methods. They are metabolic control and permeability control. The metabolic control constitutes feedback control and restriction of enzyme catalysis. In feedback control, the overproduction of nucleotides and amino acids is avoided.

Permeability

Permeability is considered as crucial in industrial microorganisms because it helps in product isolation. The permeability can be increased by some important methods. They are biotin deficiency, use of fatty acid derivatives and penicillin. Biotin acts as a coenzyme in both transcarboxylation and carboxylation reactions. This includes carbon dioxide fixation resulting in acetate and then to malonate. Biotin is rich in the enzymes catalyzing these reactions. The deficiency of biotin results in problems with fatty acid production especially long chained fatty acids. This will have a negative effect on the lipid content of the cell membrane and hence there would be a leak in the membranes. The biotin deficiency has created aberrations in, *Bacillus polymax,B. megaterium* and in yeats. The fatty acid derivatives are surface acting agents. These agents have similar importance as biotin and it is mandatory to add these fatty acid derivatives into the medium. Some of the examples of the derivatives are tween 40 or mono palmitate and tween 60 or polyoxylene sorbitan monostearate. These agents will cause changes both qualitatively and quantitatively in lipid content of the cell membrane. They enhance saturated fatty acids and limits the unsaturated fatty acids. Penicillin generally inhibits the cell wall formation. This compound interfers by cross linking acetylmuranmic and polypeptides in the mucopeptides. The cell walls due to this activity get deranged and the glutamate is excreted due to membrane damage.

Secondary metabolite regulation

The compounds have stimulatory effect which is same as the enzyme induction. The ergot alkaloid fermentation by the Claviceps sp is mediated by the tryptophan. Tryptophan is an amino acid which acts as a precursor in inducing enzymes needed for alkaloid biosynthesis. The amino acid tryptophan needs to be added to the medium especially in the growth phases. The secondary metabolism is regulated by carbon. The carbon sources used in the secondary metabolite production are starch and dextrin for production of fortimicin, glycerol for butirosin production, soyaben oil for kasugamycin production, sucrose for production of erythromycin and tetracycline. In case of catabolite repression, the enzymes required for production of secondary metabolites are repressed. Nitrogen catabolite regulation is one of the important mechanism observed in case of primary metabolism. This involves the suppression of enzyme production which acts on nitrogen containing substances such as ureases and proteases. The secondary metabolite which are affected by this nitrogen Catabolite regulation are cephamycin fermentation by *Strepromyces* spp, bikaverin fermentation by *Gibberella fujikuroi* and trihyroxytoluene production by *Aspergillus fumigates*. In these cases, given above, the nitrogen should be exhausted prior to the secondary metabolite production is initiated.

The feedback regulation generally exists in case of secondary metabolism. The biosynthesis of penicillin by *Penicillium chrysogenum* is affected by feedback inhibition presented by L-lysine as lysine and penicillin are the two end products found in the pathway. Many secondary products and their analogues have gor a tendency to inhibit their production with the help of the feedback mechanism. Some of them are cycloheximids, penicillin, chloramphenicol etc.

The secondary metabolism has low tolerance for inorganic phosphate concentrations during primary metabolism. Inorganic phosphate in the range of 0.3 to 30 mM can be used for good growth of eukaryotic and prokaryotic organisms. The concentration of 1mM favors the secondary metabolism and the concentration of 10mM suppresses the secondary metabolism. Low phosphate medium is used in the industries to prevent the inhibition of antibiotic production. Phosphate stimulates high glucose utilization, high DNA and RNA synthesis and also stimulates high respiration rates. The growth phase transfers from idiophase to trophophase. Once the phosphate is exhausted, idiophase is triggered.

The metabolic pathways which are taken up by the secondary metabolites are well studied and the approaches which disrupt the pathways which are responsible for overproduction are employed. Much work is put into understanding the primary metabolism and metabolites. The organisms which shows higher rate of productivity are generally treated with chemicals resulting in mutants. These microorganisms which are treated were randomly selected and tested for overproduction of metabolite. Addition of precursors sometimes enhances the yields in much fermentation. In production of penicillin, phenyl lactic acid is used as precursor. This precursor is extracted from corn steep liquor. In production of aflatoxin from *Aspergillus parasiticus* methionine is used. In case of mitomycin production using *Streptomyces verticillatus*, L-citruline is used as precursor. Phosphate and manganese are the two inorganic compounds which have profound effect on secondary metabolite fermentation. Manganese enhances idiophase production in case of bacillus especially in production of mycobacillin, bacitracin, subtilin, D-glutamine etc. The optimal temperature for the good rate of growth is set at 25^{0}C. In production of secondary metabolites, the optimal temperature is set at 5-10 ^{0}C.

REFERENCES

1. Agrawal, P. (1987) Biotechnol. Techn. 1, 19.
2. Almazan, O., Klibansky, M., Otero, M. A. (1981) Biotechnol. Lett. 3, 663.
3. Bertrand, J. C., Rambeloarisoa, E., Rontani, J. F., Giusti, G., Mattei, G. (1983) Biotechnol. Lett. 5, 567.

4. Bryson, V., Szybalski, W. (1952) Science 116, 45.
5. Bull, A. T. (1981) Strategies in microbial process optimization. In: Global Impacts of Applied
6. Microbiology VI, New York.
7. Bull, A. T., Ellwood, D. C., Ratledge, C. (1979) The changing scene in microbial technology. In:
8. Microbial Technology: Current State, Future Prospects, Cambridge.
9. Cho, Y. K. (1983) Biotechnol. Lett. 5, 555.
10. Ghosh, S., Klass, D. L. (1978) Process Biochem. 13, 15.
11. Gibbons, W. R., Westby, C. A., Dobbs, T. L. (1984) Biotechnol. Bioeng. 26, 1098.
12. Harder, W., Kuenen, J. G., Matin, A. (1977) J. Appl. Bact. 43, 1.
13. Harrison, D.E.F. (1978) Adv. Appl. Microbiol. 24, 129.
14. Holst, O., Hansson, L., Berg, A. C., Mattiason, B. (1985) Appl. Microbiol. Biotechnol. 23, 10.
15. Hwang, J. S., Chang, H. N. (1987) Biotechnol. Lett. 9, 237.
16. Inloes, D. S., Smith, W. J., Taylor, D. P., Cohen, N. S., Michaels, A. S., Robertson, C. R. (1983)
17. Biotechnol. Bioeng. 25, 2653.
18. Janssens, J. H., Bernard, A., Bailey, R. B. (1984) Biotechnol. Bioeng. 26, 1.
19. Jarl, K. (1969) Food Technol. 23, 1009.
20. Jensen, D. E. (1972) Biotechnol. Bioeng. 14, 647.
21. Joshi, S., Yamazaki, H. (1984) Biotechnol. Lett. 6, 797.
22. Kennedy, K. J., Van den Berg, L. (1982) Biotechnol. Lett. 4, 137.
23. Margaritis, A., Wilke, C. R. (1978) Biotechnol. Bioeng. 20, 727.
24. Monod, J. (1950) Ann. Inst. Pasteur 79, 390.
25. Moser, (1985) Imperfectly Mixed Bioreactor Systems. In: Comprehensive Biotechnology 2, New York.
26. Novick, A., Szilard, L. (1950) Science 112, 715.
27. Ottengraf, S. P. P., Van den Oever, A. H. C. (1983) Biotechnol. Bioeng. 25, 3089.
28. Rutten, R., Daugulis, A. J. (1987) Biotechnol. Lett. 9, 505.

29. Sahai, O. P., Shuler, M. L. (|984) Biotechnol. Bioeng. 26, 27.
30. Sikyta, B. (1991) Directed Selection of Microorganisms in Continuous Culture, Prague.
31. Slezfik, J., Sikyta, B. (196|) J. Biochem. Microbiol. Technol. Eng. 1,357.
32. Sortland, L. D., Wilke, C. R. (1969) Biotechnol. Bioeng. 11, 805.
33. Stieber, R. W., Gerhardt, P. (198 l a) Biotechnol. Bioeng. 23, 523.
34. Stieber, R. W., Gerhardt, P. (1981b) Biotechnol. Bioeng. 23, 535.
35. Suidan, M. T., Siekerka, G. L., Kao S. W., Pfeffer, J. T. (1983) Biotechnol. Bioeng. 25, 1581.
36. Toda, K., Ohtake, H., Asakura, T. (1986) Appl. Microbiol. Biotechnol. 24, 97.
37. Toldrfi, F., Flors, A., Lequerica, J. L., Valles, S. (1986) Appl. Microbiol. Biotechnol. 23, 336.
38. Van den Eynde, E., Vriens, L., De Cuiooper, P., Verachtert, H. (1984) Appl. Microbiol. Biotechnol. 19, 288.
39. Wimpenny, J. W. T. (1985) Microbiol. Sci. 2, 53.

CHAPTER 4

SELECTION OF MICROORGANISMS

INTRODUCTION

The industrial microbiological processes were analyzed and through this analysis it is confirmed that genetics, microbial engineering, and microbial physiology plays a major role in the enhanced productivity of microorganisms. Genetic methods have provided the way for the production of nucleotides, amino acids and nucleosides via regulatory bacterial mutants and auxotroph. Many of the animal proteins were produced by genetic manipulation of microorganisms. An ideal industrial microbe must synthesize a particular substance preferably as single product. This product must be produced in higher concentrations and the product should be excreted at the maximum rate possible or it should be accumulated inside the cells. The synthesis of a particular product should be achieved with minimal growth of cells but if the product is the microbe itself then maximum growth of cells needs to be allowed. The consumption of the substrate should be minimum so that we can achieve higher yield coefficient. The microorganisms should be able to use varying composition of raw materials and also with varied purities. There should be a minimal preprocessing of raw materials so that the cost of processing can be kept at the minimum. In case of aerobic processes, all the biosynthesis does require oxygen and it should be used at minimum since the oxygen is not a cheap substrate. The selected microbe should be resistant to bacteriophages. The microbe should exhibit tolerance to process variables such as pH, temperature and several other factors. The microbial culture must not produce any foam or any other products which jeopardize the rate of transfer of gases such as oxygen. The productive strain is constructed by keeping in view the production ability and then other traits as per the importance.

Strategies

A strategy should be designed keeping in mind several factors concerning genetic methods and their level of approaches, complexity in methods, requirements needed in their application, microorganisms and the end products. Some of the important differentiations which are to be observed are

- The acquisition or the production methods of new strains such as recombinant techniques or mutagenesis.
- The detection methods of new strains which are produced through recombination or mutagenesis which involves selection and screening methods.

Screening is nothing but the testing of all the available individuals present in the population. Selection nothing but the isolation of the variant type which is desirable under specific set of conditions which avoids redundant isolation of the individuals of the available population. The microorganism used for industrial production is constructed based on some basic knowledge of the genetic basis, metabolic pathways and the regulatory mechanisms which control the metabolic pathways. The key regulatory mechanisms are:

- Feedback inhibition of enzymes which catalyze biosynthetic pathways
- Enzyme synthesis regulation
- The factors which are involved in metabolite production

It is a good step to isolate mutants which have disrupted regulatory mechanisms and also mutants with disruption mechanisms to several mechanisms. The genetic basis of enhanced overproduction of a particular metabolite can be controlled by genetic manipulation. The identification of particular strain with certain specific properties in the available population and later isolation is a difficult task. Some of the techniques are there for the identification of particular mutants but there are no techniques which can identify and can isolate microorganisms with varying rates of overproduction. Some of the properties of mutants which are useful for industrial application. If the mutant type expresses higher rate of resistance to bactericidal or bacteriostatic compounds then they can be grown with raw materials which are contaminated by various compounds. If a mutant has the ability to use less substrate as compared to parent strain then that microorganisme can be used for best substrate utilization and so there will be high rate of economical coefficient. The mutants bring certain changes in the transport of nutrients and hence the compounds can be assimilated easily as compared to parental strains. If the mutant strain has got defects in executing anabolic pathways then there will be good accumulation of compounds such as nucleotides and amino acids. If the microbe is resistant to temperature, then the performance of that microbe at high temperatures will be good and hence saving the cost required for cooling system.

The methods used for acquisition of new strains can be classified into rational and empirical. The empirical methods were used for long time to increase the production of metabolites such as steroids, alkaloids and antibiotics via strain improvement through mutation selection. The rational methods include in vitro DNA recombination, protoplast fusion, gene manipulation, and direct visualization and enrichment techniques. These techniques use accumulation capacity of phenotypes desirable either by open continuous or closed batch cultivation. In case of direct visualization, it is possible to study the size and growth rate of the microbial colonies. The site of the transparent zone or the growth zone is studied. The protoplast and cell fusion allows yielding high number of hybrid with properties which are desirable. These methods are best suited when more number of genes is involved. DNA recombination method is used when less number of genes is involved.

Selection

The key factor in case of enrichment methods is selection pressure. In general conditions, the microorganisms which can produce certain enzymes grow. Growth medium selection is an important factor to discourage the undesirable population by altering chemical composition. In general, the media composition, cultivation procedure, sterilization method and the physical properties decides the selection pressure and it favors only certain microbe which fits well. The conditions should be strict and they should decide whether there is to be growth or no growth and also if the cultivation should allow the population growth in long time in case of continuous culture.

The selection factors can be physico-chemical factors such as pH and temperature, physical factors, inducers, corepressors, Catabolite repressors, inhibitors which can be competitive and non-competitive, ethanol, antibiotics, inhibitory substances which include heavy metals, race elements, oxygen, sulphur, phosphorus, nitrogen and carbon. The intensification of the selective pressure can be done with the combination of factors such as

- If the enzyme production is at 28^{0}C and it is hard at 37^{0}C, then the increased temperature can be a best option to intensify selection pressure.
- The nitrogen and carbon sources can repress the enzyme synthesis partially.
- If two substrates act at the same time.

These combinations can produce conditions which promote the selection and enrichment of strains which are being overproduced. The antibiotic synthesis is high in the strains which utilize the fatty acids so it is good to enrich the strains and isolate them with lipase overproduction. The enrichment methods are crucial when recombinant strains are to be isolated as detection of the strains is difficult. The microorganism can be enriched via batch culture and continuous culture.

In case of batch culture, the inoculation of microorganisms is cyclic and the microorganism can be mutagenized or non-mutagenized. Appropriate selection pressure is applied in every inoculation. Because of the continuous re-inoculations the chance of obtaining homogeneous population also increases. This is because of the dominance of microorganisms exhibiting high growth rates with rapid utilization of substrate or the resistance towards adverse conditions. The constructive strains are enriched utilizing the growth lag and it serve as an indication and marker. A strain which uses inducible enzyme synthesis uses lag phase and a strain which synthesizes enzyme constitutively is independent of lag phase. The alternating transfers induced in the media which contained repressing substrate and inducing substrate will bring enrichment gradually in the constitutive strains. It is possible to select a composition of the medium. Sterilization method. Dissolved oxygen concentration, duration of cultivation and optimal temperature and therefore determining the properties of the new strain. One of the best raw materials used in the preparation of the media especially in the tetracycline biosynthesis is corn-steep liquor. The variations in the phosphorus concentration are observed in the batches. It is evidenced by analyzing the media with decreased phosphorus concentration ranging from 20 μg/ml to 100 μg/ml. *Strepromyces aureofaciens* strains can produce good levels of antibiotics with phosphorous concentrations of 30-40 μg/ml in the medium. It is difficult to maintain the phosphorus content at constant concentrations especially in the corn steep liquor. Higher phosphorous concentrations were applied to select strains which can have the capability to produce higher rate of antibiotics even in high phosphorus concentrations. Matelova et al observed that there is an inhibition in tetracycline production due to discontinuation of the transient aeration. Multiple aeration breaks were used in regular intervals to get good amount of tetracycline concentration. Another important selection condition is the sterilization method used. Strain improvement has yielded 20-30% more production of streptomycin when shaken flasks were used but there was no improvement when large bioreactors were used with the original strain. Changed in the sterilization conditions have produced 30% more production in the strains selected in large bioreactors and there was no increase in production even when parental strains were used.

The continuous culture is the best method to enrich the strains which are overproducing. The process of enrichment can be done in turbidostat or chemostat. The growth is limited when a substrate is added in dilutions under sub-optimal concentrations in a chemostat. The microbial concentrations are kept at constant level via dilution rate in turbidostat. The content of the population varies while cultivation due to the genetic heterogeneity in microorganisms at the start of the cultivation and also depends on the mutations because of the continuous growth. If chemostat is used, the inoculation of the microbial populations into the batch cultures at specified time intervals can yield higher specific growth rates. The increased growth rate indicates the enrichment of population. In a chemostat, the selection of the microbe is based on the limiting substrate and in turbidostat excess of the substrate is used and therefore there will be different saturation constants.

The turbidostat can be employed for enrichment of microbial strains which grows at extreme pH, temperature and higher inhibitor concentrations which discourage the growth of original population. The suitable technique for accumulation of enzyme overproducing microbial strains is the chemostat. Novick, Horiuchi, Tomizawa were the first scientists to use this technique in enrichment of *E. coli* strains which can overproduce beta galactosidase. Francis and Hansche have used beta glycerophosphate to arrest the growth of *Saccharomyces cerevisiae* while using chemostat. They have selected strains after 300 generations which are grown at pH 6.0 and the strains selected were having high acid phosphatase at pH 4.2.

Downie and Garland has selected strains with trace elements. After limiting the *Candida utilis* growth in presence of chemostat by copper, the dry weight of the cells has increased multiple folds and hence changing the pathways. The cytochrome oxidase pathways were replaced by another oxidase pathway in the new strains and it is done at the level of cytochrome b. when the copper content was brought to the original concentration then the cytochrome oxidase pathway was activated. The process of enrichment in the chemostat can also be utilized for the accumulation of auxotrophs.

Isolation

The identification of the productive strains is time consuming and it is also difficult. Many methods are being developed to identify microorganisms on selective media with the help of colony morphology, rate of growth and size of the colony, colour of the colony and also clear zones around the colony. The strains can be isolated easily. The pre-screening methods can be used for the isolation of the strain types such as

- Overproducing strain type
- Auxotrophic strain type
- Strains which are resistant to organo-metal ions, metal ions, inhibitors, temperature, pH, analogues, feedback inhibition and catabolite repression
- Constitutive strain type
- Strain types with extended assimilation of substrates
- Strain types with combined properties

Overproducing strain type

This pre-screening method was proposed by Ichikawa which was used for evaluating antibacterial activity of strains which produce antibiotics. In this method, cylinders of agar with 5 mm diameter were inoculated with the mutant colonies and later the growth is allowed to take place in a thermostat with higher level of humidity. After cultivation, the production efficiency of antibiotics was done with agar plates and the test organisms. The microbial colonies which exhibited large inhibition zones than the control microorganisms are loaded into the shaking flasks. The inhibition zone size always cannot show the power of the microbe to produce antibiotics but the non-producing and low producing colonies can be discouraged. The pre-screening for enzyme producers depends on whether exoenzymes or endoenzymes produced. In case of endoenzymes producers, staining techniques, permeabilization, and colony stain intensity techniques can be used. The bacterial colonies which produce beta galactosidase can successfully stained by o-nitrophenyyl-beta-D-galactoside and also with fluorescein-di-beta-D-galactopyranoside and napthol-AS-Bi-beta-D-galactopyranoside. O-nitrophenylphosphate can be used in assay for phosphatase activity. In case of exoenzymes producer, the zone is tested using colour indicator and the clear zone formation around the microbial colony. In order to detect amylase, the compound named starch azure is added to the medium and this compound will give blue colour to the zone due to enzyme activity. In case of beta glucosidases, a black precipitate zone is observed and this is due to the action of iron salts with esculetin. In case of inulases and pectinases, a clear zone appers, this is due to the hydrolysis of pectin and dissolution of inulin. The action of polygalacturonic acid transeliminase is detected by spraying hydrochloric acid and after that clear zone will be seen around the microbial colonies.

The auxotrophic strains are important in production of nucleosides, nucleotides, amino acids and also used in microbiological and genetics assays. These strains can be isolated by penicillin overlay method and also replica method. The strains which are resistant to chemical and physical factors can be detected on the solid media. The solid media is enriched with inhibitory and toxic substances so that only the strain resistant to these substances grow.

Industrial microorganisms which are selected through mutation selection improvement technique is capable of producing high rate of secondary metabolites. The mutation selection improvement basically depends on population which is genetically different. This is because of the effect of mutagen on the strain. This step is followed by selecting strains with better properties from the population which is mutagenized. The strains selected are once again subjected to mutagenic intervention and the process is done several times. This step of mutation selection helps in enhancing the ability to produce in industrial microorganisms. Optimization is the key principle which is attained by decreasing the time which is essential for the individual cycles. The selection and mutagenesis depends on the empirical rules and least dependent on the theoretical genetical knowledge. The mechanism of individual mutagens at the molecular level is well established. Alkylating agents such as N-methyl-N-nitro-N-nitrosoguanidine, ethyleneimine, nitrogen mustard gas was used successfully. Ethyl methyl sulphonate has not attained any recognizable effect. The selection of appropriate mutagen is based on the strain selected. Some of the mutagens which yielded successful results are UV-light plus ethyleneimine and UV-irradiation plus nitrogen mustard. The same mutagen is used repeatedly often resulting in good result but the mutagen can be changed for better results. After selection of appropriate mutagens, the other factor which needs to be considered is the strain selection and the number of strains. The start of the mutation selection can start with a single strain and can give high yields. Some of the prerequisites for a good mutation selection process are

- Appropriate system to obtain mutant
- Screening system which is reliable
- Good evaluation method for testing results
- Analytical methods which are reliable
- Good storage and preservation of cultures

The instability of strain, cultivation yield variability and the errors present in the assays will exhibit variations in results. This condition can be mitigated by several tests to get a new strain. The screening system is the key to select isolated which can yield high production rates. The selection procedure can be optimized and it includes

- Determination of the number of steps involved in screening
- Determination of the number of isolates which are to be tested in the first screening
- Determination of the number of strains which are to be taken for screening at start of each cycle.

The statistical evaluation of screening procedures has proved that the two-stage screening is the best method. The strains which are selected at the start of cycle are based on frequency of strains which gives higher production. The strains which are selected are to be tested on the pilot scale with conditions which are identical to the large-scale production set up. The improvement is done by strain selection which best suits for cultivation conditions and then the medium selection. Sometimes the strains which give better yields on a pilot scale or small scale cannot yield much in large scale production. The best process for identifying the failure of strain is by testing via scale down process.

Mutasynthesis

Mutasynthesis involves an antibiotic producing-microorganism which is modified to get a defective microorganism in which a step in its biosynthetic pathway which is responsible for a portion of the antibiotic molecule is blocked.. The mutant whose biosynthetic pathway is blocked is supplemented with artificial or natural precursors and they produce antibiotics. The cells should have a capacity to accumulate precursor.

In case the cells are unable to assimilate the precursors, then in vitro systems can be used. The defective mutants or antibiotic negative are called as idiotrophic. The process of mutant biosynthesis was first employed by Gottlied, Rinehart and Shier in *Streptomyces fradiae* strains which can produce neomycin A, B and C. This strain is used to produce diaminocyclitol part of deoxystreptamine. Mutagenesis produces a mutant which cannot biosynthesize deoxystreptamine. This disrupted biosynthesis is restored by adding deoxystreptamine to the medium. When the strain is cultivated in the medium containing streptamine, the mutant strain has produced hybrimycin A1 and A2, hybrimycin B1 and B2, C2 primer of streptamine and Epistreptamine. The analogues of butirosin were produced by the process of mutasynthesis using Bacillus circulans which is a mutant strain. The process of mutational biosynthesis is used for production of secondary metabolites but the antibiotic production was not reported. By the process of mutation biosynthesis, *Pseudomonas aureofaiens* produced 4-fluoropyrrolnitrine instead of Pyrrolnitrine, *Streptomyces peuceticus* produced adriamycin instead of Daunomycin, *Nocardia mediterranei* produced rifamycin W instead of Rifamycin B, *Micromonospora inyoensis* produced mutamycins instead of Sisomycin, *Streptomycin griseus* produced streptomutin instead of streptomycin, *Acremonium chrysogenium* produced 6-D-2-amino-2 Carboxy ethylthio acetamido penicillanic acid instead of Penicillin N and *Micromonospora purpurea* produced hydroxyl and deoxy gentamycin instead of Gentamycin.

Deregulation of pathways

The regulatory mechanisms control the overproduction of primary and secondary metabolites. If the metabolite overproduction is needed, then the regulatory mechanisms need to be disorganized. *Cornyebacterium glutamicum* produced Lysine and it exhibits auxotrophy for threonine. *Corynebacterium glutamicum* exhibits auxotrophy for methionine and threonine, and homoserine and it produces lysine. *Brevibacterium flavum* exhibit auxotrophy for lysine and it produced diaminopimelic acid. The mutants comprise the base for the amino acid biosynthesis, nucleosides and nucleotides and play a crucial role in production of steroids, antibiotics and enzymes.

Amino acid overproduction

The deregulation in the metabolic pathways can be illustrated by amino acid overproduction. The amino acid overproduction by auxotrophic mutants and they are defective in repression mechanism or feedback inhibition. The amino acid overproduction can be attained by regulatory mutants also. The overproduction process in this case id the avoidance of feedback regulatory mechanisms or removing sensitivity of mutant towards feedback regulation. The mutants obtained are stable and not prone to reverse mutations.

Nucleotide overproduction

The deregulation of pathways trough isolation of regulatory and auxotrophic mutants can attain nucleotide overproduction. The strains which produce inosine in higher rate can be arrested in adenyl synthetases to stop the regulation by phospohoribosylpyrophosphate amidotransferase on nucleosties. Inoisne is arrested in nucleosidase to stop inosine decomposition to hypoxanthine. Some of the important conditions which are to be followed while acquiring strains which are capable of overproducing guanine are

- Deficiency of guanosine monophosphate reductase and adenylosuccinate synthetases
- Lowering of hydrolysis rate of purine nucleosides
- Enzyme deregulation of GMP pathway especially phosphoribosylpyrophosphate amidotransferase.

Metabolite overproduction

Bacillus licheniformis strains produce antibiotic bacitracin. The limiting factor in production non-protein amino acid is ornithine. A block at the gene level for ornithine degradation in presence of ornithine beta transaminase can enhance the bacitracin production and lowering the proline and glutamate concentrations in the cells. Auxotrophic mutants, antimetabolites, metabolite analogues are used for the metabolite overproduction. Methionine involves in chlortetracycline biosynthesis and the reversal of mutant auxotrophs to primary metabolite can be a good tool for secondary metabolite overproduction. The pyrrolnitrine which is an antifungal antibiotic was overproduced in mutants which are resistsnt to methyltryptophan and fluorotryptophan. The penicillin and cephalosporin were overproduced in the presence of precursor's valine, cysteine and alpha aminoadipic acid while using mutants. The strains which are sensitive to antimetabolites and growth inhibitors overproduce cephalosporin C. The growth inhibitors in this case can be p-fluorophenylalanine and acriflavine. Cephalosporium acremonium is sensitive to selenomethionine and it concumes exces of methionine to produce Cephalosporin C. In *Candida lipolytica*, Monofloroacetate inhibites aconitate hydratase which catalyzes the conversion of citrate to isocitrate. The mutants which are sensitive to monofluoroacetate can produce lower amount of aconitate hydratase. The heavy metals like copper and mercury can produce complexes which are crucial in beta lactams antibiotics production. The mutants which are resistant to metal ions can produce beta lactams in higher concentrations and so there is metal detoxification. The strains of *Streptoomyces lipmanii* which exhibit copper phenylacetate can produce higher amounts of cephamycin. This strain when resistant to mercury chloride can produce higher amounts of cephalosporin C.

Fusion of cell and protoplast

The microorganisms via the process of natural recombination can produce recombinants but the rate of recombinant population is low. The cell wall acts as a barrier to stop the fusion of cells even between the same species. The recombinants can be allowed to fuse their cell contents with each other can be achieved by the cell wall breakage. In this method, the cell fusion is induced and in some cases if cells cannot fuse, the protoplast are allowed to fuse. The cells or the protoplasts can be transformed by the DNA via fragments, phage vectors and plasmids. In this technique, the crossing can be intergeneric, intraspecific and interspecific.

Cell fusion

The genetic material can be transferred to the recipient cells when the cells were treated with acetates or lithium acetates and these substances acts on the cell surface. The other method which is important in the transferring the genetic material is the microinjection. A borosilicate glass micropipette or the micromanipulator is used to inject recipient cells with protoplasts or the cell contents, mRNA, DNA, chromosomes and so on. Electroporation or the electrofusion is another important technique used for the transfer of cell and protoplast contents.. In this process, the electric current is utilized to form pores in the cell wall and these pores allows fusion of cell contents. The electroporation device comprises of reactor and regulator. This device controls the electric field and encourages cell-cell interaction. The parameters like duration and number, frequency and pulse amplitude can be adjusted precisely. The cell to cell interaction generally takes place in a reactor. This reactor contains electrodes which are positively charged. The distance between the electrodes can either be decreased or increased to encourage contact free action or contact action. The reactor vessel can accommodate a volume of 1ml and the entire treatment process takes three minutes or often less.

Fusion of protoplasts

The important factors to be considered while acquiring protoplast are the cell wall composition and the physiology of the cells or the age of the cells. In case of gram positive prokaryotes like *Bacillus*, *Streptomyces*, *Micrococcus* and several others, the protoplasts are generally prepared using lysozyme on the cells. The lysozyme concentration which is needed for protoplast preparation in *Micrococcus luteus* will be 1microgram/ liter and the concentration will be 50 micrograms /liter in *Bacillus megaterium*. In many microorganisms, the protoplast formation is done by adding subinhibitory substances such as glycine into the nutrient medium. The lysis of the gram-negative cells generally takes place in presence of EDTA in the medium. The cells are dissolved in hypertonic sucrose solutions preferably with 0.1M or 0.2 M concentrations in order to achieve stable protoplasts. The protoplasts in fungal members can be obtained by the help of enzymes which are capable to break gucan manan complex or chitin complex. The protoplasts of yeast can be obtained by treating the cells with glusulase, sulphatase, helicase which are basically lytic enzymes present in the gut of snail (*Helix pomatia*) or by the help of zymolyase enzyme which is obtained from *Arthrobacter luteus*. In case of filamentous fungi, the process is complicated. The enzyme mixtures comprised of lytic complexes, cellulases, chitinases, glusulase are generally used for the protoplast preparation in the filamentous fungi. The protoplasts present in the spores are obtained by the prolonged treatment using lytic agents. The fungal protoplasts can be stabilized using saccharides in case of yeast and organic salts such as magnesium sulphate in the concentration of 0.6 M in case of filamentous fungi. The protoplast regeneration can be done by washing off the protoplasts onto the agar medium which is osmotically stabilized or via direct plating onto the agar medium. The protoplasts from the filamentous fungi can be regenerated using liquid media. The reversion frequencies always vary.

The protoplast fusion is done in optimal conditions and by using polyethylene glycol (PEG). It was first employed by Kao et al. (1974) with plant cells. PEG is used with plant, animal cells and other microbial cells. The concentrations of PEG will vary and it is species specific. In case of bacteria the optimal concentration will be 30-50 %(w/v). The fusion of the protoplast can be stimulated by dimethyl sulphoxide and calcium. The exposure time of protoplasts to PEG is less than one minute. The PEG concentration in case of fungal protoplast fusion is less when compared with animal and bacteria cells. The optimal range of PEG will be 25-30%. Results are good when higher concentrations of PEG are used. The optimum pH for the protoplast fusion is 9 and this is in presence of calcium which enhances the efficiency of the fusion. The dead donor technique is used in prokaryote microorganisms in order to obtain high number of recombinants. The protoplast of one strain is primarily exposed with streptomycin or lethal heat dose or to light as in *Streptomyces*. Both the stains can be inactivated. The genetical work which is based on the protoplast fusion can also occur naturally and it is the best method to identify the recombinant strains. The traits which are to be considered includes resistance and auxotrophy or the mixture of both. The acriflavine resisance and auxotrophy in case of *Aspergillus rugulosus* and *Aspergillus nidulans* can be used as identification markers instead of acriflavine sensitivity and prototrophy.

Applications

If the direct transfer of the DNA into the competent cells is difficult, then the best alternative is protoplast fusion. The methods which involve protoplast fusion, plasmids, transformation and transduction often result in strains which overproduce enzymes. Fusion of two different strains of *Brevibacterium glutamicum* with one strain which produced high amount of lysine with slow growth rate and the other strain which produced low amount of lysine with higher growth rates produced recombinant strains with lysine overproduction. In case of *Cephalosporium acremonium*, low producing and high sporulating strains were fused with non sporulating strain which required inorganic sulphate. The recombinant strain which is generated can grow at higher rate and also produce cephalosporin with the help of sulphate. The production of antibiotic resulted in an increase of 40 percent.

The protoplast fusion was also used with mutagenesis procedures. The diploid crosses were resulted while fusing *Candida tropicalis* with *Saccharomyces fibuligera* and *Saccharomyces cerevisiae* with *Schizosaccharomyces pombe*. These organisms comprised all the enzymes of parents. The frequency of the fusion is low and the isolates produced exhibited different stabilities. *Streptomyces kanamyceticus* which produced kanamycin fused with *Streptomyces rimosus* v. *kanamyceticus* which produced kanamycin produced recombinants which produced neomycin.

Recombination of DNA

In case of DNA recombination, small portions of genes will be participated in the synthesis of product. A laboratory technique on a small scale will be sufficient to use restriction enzymes in order to get DNA fragments and later joined with the help of ligases. The construct is allowed to transfer into the protoplast or the cell with the help of suitable vectors. Some of the steps involved in the procedure are

- Enzyme selection
- DNA selection
- Vector selection
- Recombination of donor DNA and vector DNA
- Cell transformation
- Detecting cells with recombinant DNA
- DNA amplification
- Modification of gene

- Gene modification and insertion to suitable expression vector
- Transformation of cells
- Cell growth, isolation of plasmids and their characterization
- Cell growth with a desired product

Cloning

The donor DNA can be from plant, human cells, animals and microorganism. The DNA strands are naturally fragmented during the isolation process and shorter fragments are achieved by the help of restriction endonucleases. The restriction enzymes which are generally used possess hexanuclotide recognition site and they give DNA fragments which contain four base single strand termini. The frequency of each sequence in any polynucleotide which possesses 4 bases in equal proportions is 1 in 4^6. These DNA fragments are injected into a vector DNA and it constitutes as one of the component of vector/host system. The host cell can be eukaryotic cell or prokariotic cell. A good vector is a small DNA fragment which has the capacity to replicate autonomously when present in the host cell. Based on the enzyme utilized in the donor DNA fragmentation and later vector DNA linearization, the DNA molecules should contain single strand ends which are protruding and are with complementary nucleotide sequences. This helps spontaneous yielding of recombinant DNA. DNA ligase will close the DNA strands covalently. The host cells which are suitable will take the vector DNA along with the donor DNA by the process of transformation. The transformed cells get reproduced and the recombinant DNA in vitro will have the capacity to replicate autonomously and therefore yielding recombinant DNA clone.

Vectors

The vectors which are selected for the cloning process should be small, genetically and physically well defined and it should be easily purified. The DNA fragment insertion should not deactivate the genes present in the vector DNA. The vector DNA needs to have many target sites which are meant for restriction enzymes and other DNA fragments and it should be detected easily. It needs to contain certain strong promoters in the vicinity of the insertion site to make sure that the foreign DNA fragment transcription is possible. The detection depends on the resistance conferred by antibiotics. Some of the vectors include shuttle vectors, cosmid, phage and plasmid.

The plasmids are extrachromosomal DNA particles which are covalently closed and they carry varied number of genes ranging from 2 to 250. They are present in the cell cytoplasm and sometimes gets integrated into the chromosome. The number of copies is generally inversely proportional to the plasmid size. The plasmid vectors were manufactured from the microorganisms namely from the genera *Saccharomyces*, *Strptomyces* and *Escherichia*.

The phage vectors are taken from Ø C31 actinophage of *Strepromyces lividans* and *Escherichia coli* bacteriophages. They contain some segments which carry genes and they have no connection with phage DNA replication. The segments were deleted to accommodate target sequences which are needed for restriction enzymes. The DNA fragment which is to be inserted must have the same molecular weight as the DNA segment which is deleted. The phage vectors permit the cloning of foreign DNA fragments which arc large. The phage vectors are having certain advantages such as chromosome integration, acquisition of large cloned DNA fragments and the phage induction.

The cosmid vectors were used for large segment cloning of ecukaryotic and prokaryotic genomes in the *E. coli*. Cosmids contain plasmid DNA with markers and cohesive terminal sites or cos sites from lambda phage sequences. The plasmid molecules have cos sequences which can be recognized by lambda proteins and also protein precursors. The cloned fragments range around 4kb. The receptor sites present on the cell surface are used for the cosmid molecule penetration. The hybrid vectors which contained plasmid and lambda phage allows the cloning of larger DNA fragments and they are used in acquisition of genome libraries. The shuttle vectors are constructed to study gene expression when heterologous hosts were used and also used in intergeneric gene transfer. They are constructed by the union of two vectors namely *Escherichia-Saccharomyces*, *Escherichia-Streptomyces* and *Escherichia-Bacillus*.. Some of other plasmids used are pIJ101 and pBR322. pAM1 is a vector which mainly consists of *E. coli* DNA and it carries leu^+ gene which serve as an eukaryotic marker. The pCP2 vector consists of a fragment of mitochondrial DNA which is from *Cephalosporium*. It is advantageous to use molecular cloning using eukaryotes along with hybrid vectors. The hybrid vectors basically contain eubacterial segment and this helps in fast replication when present in eukaryotes. These types of vectors are used both in eukaryotes and prokaryotes.

Drawbacks of molecular genetics methods

There are some of the problems which need to be addresses while using molecular genetics methods especially in the industrial processes. Some of the problems are instability in the extra chromosomal elements, toxicity of products, and incomplete expertise in knowing genetic basis of the industrially important microorganisms. The genetic instability which is observed in the extra chromosomal elements does create some difficulties. The recombined plasmids are prone to lost because of the several mechanisms involved such as deletion, recombination with chromosome and segregation. The genetic instability can be lessened by the following methods.

1. Insertion of gene of interest into the plasmid like
    a) Resistant gene to an particular antibiotic which is introduced into host
    b) A gene which encodes a phage repressor is introduced into a host which is lysogenic.
    c) A gene which is functional is used for DNA synthesis can be introduced into the host which is mutant for this gene
2. Regulation of cloned genes expression
3. The utilization of vector replicons
4. Homology elimination between host chromosome and extra chromosomal elements by utilizing host cells which are recombinant and defective.
5. Avoiding the extra chromosomal elements which have the capability to transfer transposons

The method which used antibiotic resistant gene incorporation into the plasmids followed by the introduction into the cell is a good method for plasmid maintenance in the cells. But this method is not feasible economically with bulk productions which utilize microorganisms. It is good in the production of expensive products such as enzymes, hormones and interferons. Lysogenic phage may carry repressor mutation; the plasmid loss can result in cell death. The cells can be phage resistant till they carry plasmid which contains repressor genes. The expression of the clone genes helps in maintenance of the plasmid. The metabolic stress can be detected at the growth phase or at the end of it. This method reduces the cell growth when plasmid is lost.

Some of the approaches which needs to be known are

a) Molecular genetics techniques need to be used when isolating genes and related elements from the microorganisms which are industrially important.

b) Systems needs to be developed to make a gene transfer into fungi or bacteria which are industrially important

c) Genetic systems in the microorganisms which are industrially important needs to be developed and the utilization of their own genetic elements in the form of vector is a good practice.

The toxicity of the product can be regulated by

1) Gene transfer into the microorganism which is tolerant to the product

2) Gene transfer to the microorganisms meant for production

3) Molecular basis of toxicity and its tolerance with respect to certain product needs to be determined.

REFERENCES

1. Akiyama, S., Suzuki, T., Sumino, Y., Nakao, Y., Fukusa, H. (1972) Agric. BioL Chem. 36, 339.
2. Arber, W. (1985) Swiss Biotechnol. 3(1), 7.
3. Bialy, H. (1986) Biotechnology 4, 683.
4. Broach, J. R. (1981) The yeast plasmid 2 p circle. In: The Molecular Biology of the Yeast

5. Saccharomyces. The Life Cycle and Inheritance, Cold Spring Harbor.
6. Brown, S. W., Oliver, S. G. (1982) Eur. J. Appl. Microbiol. Biotechnol. 16, 119.
7. Calam, C. T. (1964) Progress Ind. Microbiol. 5, 1.
8. Carter, B. L. A., Doel, S., Goodey, A. R., Piggott, J. R., Watson, M. E. W. (1986) Microbiol. Sci. 3, 23.
9. Celis, J. E., Graesmann, A., Loyter, A. (1980) Transfer of Cell Constituents into Eukaryotic Cells.
10. New York.
11. Chang, L. T., Elander, R. P. (1979) Dev. Ind. Microbiol. 20, 367.
12. Churchward, G., Chandler, M. (1985) In vitro recombinant DNA technology. In: Comprehensive
13. Biotechnology 1, New York.
14. Cohen-Bazire, G., Jolet, M. (1953) Ann. Inst. Pasteur84, 937.
15. Cuskey, S. M., Frein, E. M., Montenecourt, B. S., Eveleigh, D. E. (1982) Overproduction of cellulase
16. -- screening and selection. In: Overproduction of Microbial Products, London.
17. Dalbadie-McFarland, G., Cohen, L. W., Riggs, A. D., Morin, C., Itakuba, K., Richards, J. H.
18. (1982) Proc. Nat. Acad. Sci. USA 79, 6409.
19. Daum, S. J., Lemke, J. R. (1979) Ann. Rev. Microbiol. 33, 241.
20. Davis, B. D. (1948) J. Am. chem. Soc. 70, 4267.
21. Demain, A. L. (1971 a) Methods Enzymol. 22, 86.
22. Demain, A. L. (1971b) Adv. Biochem. Eng. 1, 113.
23. Downie, J. A., Garland, P. B. (1973) Biochem. J. 134, 1051.
24. Dulaney, E. L., Dulaney, D. D. (1967) Trans. N.Y. Acad. Sci. 29, 782.
25. Elander, R. P. (1982) Traditional versus current approaches to the genetic improvement of microbial
26. strains. In: Overproduction of Microbial Products, London.
27. Elander, R. P., Chang, L. T. (1979) Microbial culture selection. In:

Microbial Technology 2, New York.

28. Elander, R. P., Mabe, J. A., Hamill, R. L., Gorman, M. (1971) Folia Mierobiol. 16, 157.
29. Englesberg, E., Irr, J., Power, J., Lee, N. (1965) J. Bacteriol. 90, 443.
30. Esser, K., Oeser, B. (1985) Process Biochem. 20(3), 86.
31. Fodor, K., Demiri, E., Alf61di, L. (1978) J. Baeteriol. 135, 68.
32. Francis, C. J., Hansche, P. E. (1972) Genetics 70, 59.
33. Gerbaud, C., Fournier, P., Blanc, H., Aigle, M., Heslot, H., Guerineau, M. (1979) Gene 5, 233.
34. Godfrey, O. W. (1973) Antimicrob. Agents Chemother. 4, 73.
35. Grosjean, H., Fiers, W. (1982) Gene 18, 199.
36. Hamlyn, P. F., Ball, C. (1979) Recombination studies with Cephalosporium acreminium. In: Genetics
37. of Industrial Microorganisms, Washington, D.C.
38. Hardman, D. J., Cowland, P. C. (1985a) Microbiol. Sci. 2, 90.
39. Hardman, D. J., Cowland, P. C. (1985b) Microbiol. Sci. 2, 184.
40. Herschberger, C. L., Larson, J. L., Fishman, S. E. (1983) Ann. N. Y., Acad. Sci. 413, 31.
41. Hirose Y., Enei, H., Shibai, H. (1979) Ann. Rep. Ferm. Proc. 3, 253.
42. Hopwood, D. A. (1970) The isolation of mutants. In: Methods Microbiol. 3A, London.
43. Hopwood, D. A. (1981) Ann. Rev. Microbiol. 35, 237.
44. Hopwood, D. A., Wright, H. M. (1979) J. Gen. Microbioi. 111, 137.
45. Hopwood, D. A., Wright, H. M. (1981) J. Gen. Mierobiol. 126, 21.
46. Hopwood, D. A., Bibb, M. J., Bruton, C. J., Chater, K. F., Feitelson, J. S., Gil, J. A. (1983) Trends
47. Biotechnol. 1(2), 40.
48. Horuichi, T., Tomizawa, J. I., Novick, A. (1962) Biochim. Biophys. Acta 55, 152.
49. Hsu, E. J., Vaughn, R. H. (1969) J. Bacteriol. 98, 172.
50. Ichikawa, T., Date, M., Ishikura, T., Ozaki, A. (1971) Folia Microbiol. 16, 218.

51. Jackson, D. A. (1981) Molecular genetics and microbial fermentations. In: Trends in the Biology
52. of Fermentations for Fuels and Chemicals, New York.
53. Jones, D. T., Woods, D. R. (1986) Microbiol. Sci. 3, 19.
54. Kao, K. N., Constabel, F., Michayluk, M. R., Gamborg, O. L. (1974) Planta 120, 215.
55. Khorana, H. G., Agarwall, K. L., Besmer, P., Btichi, H., Gruthers, M. H., Cashion, P. J., Fridkin,
56. M., Jay, E., Kleppe, K., Kleppe, R., Kuma, A., Loewen, P. C., Miller, R. C., Minamoto, K.,
57. Panet, A., Ray Bhandary, U. L., Ramamoorty, B., Sekiya, T., Takeya, T., van de Sande, J. J.
58. (1976) J. Biol. Chem. 251,565.
59. Kielland-Brandt, M. C., Casey, G. P., Gjermansen, C., Holmberg, S., Nilsson-Tillgren, T., Petersen,
60. J. G. L., Polaina, J. (1986) Swiss Biotechnol. 4(1), 18.
61. Kinoshita, S., Nakayama, K., Udaka, S. (1957) J. Gen. Appl. Microbiol. 3, 276.
62. Krumphanzl, V., Sikyta, B., Van6k, Z. (1982) Overproduction of Microbial Products, London.
63. Leuchtenberger, W., Kircher, M. (1986) Swiss Biotechnol. 4(1), 18.
64. Levi, C., Sanches Rivas, C., Schaeffer, P. (1977) FEMS Letters 2, 323.

CHAPTER 5

ESSENTIAL MATERIALS

INTRODUCTION

The raw materials used which are used in microbial production are of huge demand when compared with other industries. There will be no compromise in maintaining of the quality of the raw materials because the end products obtained are to be used in the food industry and medicine. The microbial growth is most sensitive to the trace substances which are present in the raw materials. The raw materials are utilized for the nutrient media preparation, product isolation and the processing till the end product is achieved. The nutrient media in microbial growth mimics the chemical medium while progreesing a reaction. The nutrient media must satisfy the demands which allow the biochemical processes. The medium should allow the growth of the microorganisms and the formation of products. The medium is produced by transforming the raw materials or the substrates into a liquid form but some substances which gives liquid nutrient medium can be gaseous or solid substrates. The substrates can be classified into basic, macroergic and intermediate. The basic substartes generally comprise of molecules with low chemical energy such as nitrogen, carbon dioxide and water. The macroergic substrates contain high energy molecules. The intermediate substrates form when the basic substrates gets converted to high energy substrates.

The choice of raw materials is critical and it is a crucial step to decide whether to go for a small-scale production of expensive products or large-scale production of cheaper products. The raw materials are selected through careful consideration of final product quality, water and energy requirements, stability, and processing before utilization. These considerations are essential as they will decide the investment costs and operation costs and hence increase in product price. The industrial microbial processes are prone to pressure for getting better and cheaper substrates. Cheaper substrate can be any substrate which is having low stability and pose problems with the sterilization. The main advantage of the microbial process is that there exists an opportunity to utilize various raw materials so that if one raw material price increases then the other cheap raw material adjusts the whole price. In some cases, all the raw materials may suffer with high prices, in that scenario it is good to increase the price of the end product. The shift in product range if imminent due to rapid variations in their prices. When continuous increase of prices of raw materials is seen, then there would be an enhanced product cost and it will be at critical level making it inacessable to the users. Then the user shifts to chemical processes if the microbial processes are costlier relatively. A considerable change in the raw materials may force the producer to shift to alternative technology. When formulating a formula for preparation of nutrient media, the cost of carbon resources is considered since they are required in heavy quantities. The raw materials which are to be used in the microbial processes needs to be avaialable all the year. Some of the raw materials are available in particular seasons and there will be contamination problem during their maintenance. The other important factor which is to be considered is the transportation cost. It is a good practice to install a prodction plnat nearer to the resource point location.

Water

The main problem with the enzymes is that most of the reactions occur in aqueous medium. In microbial processes, the reactants are ket at low concentrations. The reasons for this condition is

- The microbial catalysts are maintained at lower concentrations because the growth of microorganisms is limited by oxygen and various substrates.
- The products of the reaction acts as the production inhibitors and they are generally present in lower concentrations.

- If the substrates are maintained in higher concentratins then there would be product inhibition.

In order to maintain good and feasible economics, it is essential to maintain low concentrations of enzymes and substrates. Most of the enzymes requires aqueous medium but there are certain enzymes which requires liquid-liquid interface or higher concentrations of substrate. Though many advances exist in microbial production, the product concentration is low. If the products are amino acids and organic acids, then the product percentage is around 10 to 12%. If the product is antibiotic the product percentage is 3% and in case of vitamins the product percentage decreases even to 1%. Large volumes of water are used for washing, cooling and auxillary operations. It would be a good step to install water recirculation mechanism especially in large scale plants. Nutrient media preparation always demands drinking water so as the washing of microbial biomass in industrial microbial processes. Technical quality water is generally utilized in cooling of media post sterilization for cooling and regulating temperature of the culture. This water is also used in washing isolation and cultivation devices. If the drinking water is scarce, then the processed water which is relatively of low quality can be used. The low-quality water can be treated with the following steps. They are

- Chemical treatment for the removal of iron, free chlorine stripping, carbonic acid elimination and water hardness removal.
- Filtration and sedimentation to remove suspended particles.
- Disinfection, sterilization, and filtration for microbe removal.
- Use of reverse osmosis for colloid separation

The water quality should be checked regularly by biological and chemical tests. If water contains traces of nitric acid salts, nitrous salts or ammonia with concentrations ranging at 100 g per 1 ml is not accepted. If water contains heavy metals, then there will be growth inhibition and corrosion is observed in the cultivation devices. Magnesium and calcium concentrations can create complications in absorption of nutrients by microorganisms.

Raw materials (plant)

The raw materials include forestry or agriculture such as wood pulp, tuber crops, root or cereals and others. These materials are classified into different groups. They are primary or major materials, secondary or auxiliary materials and waste materials. The primary materials are lignocellulosic materials, plant oils, polysaccharides, and saccharides. The secondary materials are corn steep liquor, molasses, and others. The waste materials can constitute of various compositions. The water content is higher in waste of vegetable and fruit processing, potato tuber wash and sulphite liquors and lower water quantity is observed in case of seed husks and soil plant waste. The difference between auxiliary and waste materials cannot be easily observed. The waste materials are reused for the microbial processes. The Biological methods such as fermentation, enzymic hydrolysis or chemical hydrolysis, drying and evaporation, physical and chemical methods are used to process the auxiliary and raw waste materials. The plant raw materials can be advantageous as they are renewable but some of the plants demands higher energy costs.

Mono and oligosaccharides

D-glucose and sucrose are the traditional carbon sources for the growth of microorganisms. The auxiliary materials constitute sugars such as wood pulp prehydrolysates, date syrup, hydrol, cane molasses and beet molasses. The waste materials related are sulphate liquors which are produced from the processing of plant products especially in the food industry. D-glucose is predominantly found in trees and bushes and larger amounts are present in grapes which measures at 10 to 30 weight percent. This compound is industrially produced by the process of enzymic hydrolysis or acid hydrolysis of starch. D-Glucose can also be produced by the process of hydrolysis of cellulose molecules extracted from the lignocellulosic compounds. D-Glucose is costlier than sucrose. Precaution should be maintained while sterilizing glucose due to the production of brown substances especially in the higher temperatures. These brown products are formed when glucose reacts with phosphates, ammonium ions, amino acids, and proteins. These products cannot be assimilated by the microorganisms. Because of these limitations, hydrol is used instead of glucose. Hydrol is formed as a byproduct during the glucose production from starch. It is generally extracted in its solid form and it constitutes around 50 weight% of glucose.

Sucrose is generally produced from sugar cane and beet in the sugar factories. The degree of purity varies in case of molasses, raw sugar, and refined sugar. The degree of purity is important in deciding the use of the sucrose especially as solvents, alcohols, biomass, and antibiotics. The price of glucose varies in different countries. Sucrose production is abundant in Caribbean and Central America and it is lease produced in the USA.

Molasses is being used as a substrate in various grades. The molasses with low grade is used in production of ethanol and citric acid. High grade molasses is used in production of antibiotics and baker's yeast. Molasses is thick and dark in color with a syrupy texture. Acid molasses is not considered for the microbial processes because it contains higher percentage of invert sugar, volatile acids, nitrites and sulphur dioxide. If the molasses contain 10^6 microorganisms/g then it is not suitable.

Date syrup is a viscous fluid and it is brown in color. It is produced by heating dates in presence of water and later filtration followed by concentration. It contains sulphates and phosphate and possesses low nitrogen percentage. Citric acid production is good when date syrup is used rather than molasses. The processing of vegetables and fruits produce large amounts of waste. Apple pomace is used for production of acetone-butanol, ethanol and biomass. Mushes from bananas, mangoes, berries, forest fruits. Peaches and apricots can also be used. Vegetable wastes which can be included are onion remnants, legumes and beans which are boiled, liquors and sauerkraut juice. The main limitation is that they do contain lower percent of carbon source.

Polysaccharides

The polysaccharides are one of the best carbon resources which can be used for microbial growth as they are easily available and can be readily utilized by the microorganisms. The polysaccharides which are mostly used are starch, inulin, and cellulose. These substances are polymeric and they are capable of forming colloids and are not soluble in water. These substances need to be transformed into their soluble forms before using as ingredient in nutrient media.

Starch

Starch is generally present in plants and they are capable to form granules and can be deposited in leaves, tubers, roots and seeds. The name of the starch is based on their origin such as potato starch, barley, maize and others. The starch grains are hygroscopic in nature. In cold temperatures, the starch grains remain unaltered. If starch grains are subjected to higher temperatures and if moisture is present then they swell and form starch paste. This reaction occurs when water is present in certain ratio such as 1:4 at 100 ^{0}C. If the temperatures are maintained between 120 ^{0}C to 130^{0}C, the paste can be liquefied and it can be enzymatically or chemically hydrolyzed. Starch is made up of 20 % w/w amylose which is water soluble and when subjected to iodine treatment in cold temperatures, it stains blue. It also contains 80% w/w amylopectin and this substance forms a paste in presence of warm water and when subjected to iodine treatment it stains violet. Amyloses are made up of α glucopyransose and form lengthy unbranched chains. One molecule of amylose contains 1000 glucose residues and the molecular weight varies between 4 X 10^3 to 1.5 X 10^3. Amylopectin is made up of 3000 glucose units and the molecular weight is 5 X 10^5. These molecules are also branched. Starch is extracted from three plant types. They are tuber crops such as potatoes, cereals such as maize and tropical plants such as manioc and palms. The starch content in plants is generally 75% in rice and 19% in potatoes. High rate of starch is produced from Cassava tapioca or manioc and from Manihot esculenta or Glantz which are grown from Argentina to the USA. This plant possesses higher rate of photosynthesis. The roots of this plant constitute 35% starch. The crop yield 13 tonnes per year in Brazil in natural soils which are untreated and 60 tonnes per year if culture plots are used. Manioc contains an appreciable amount of hydrogen cyanide and this can be removed by washing or drying. In hybrid varieties of Manioc, there is little amount of hydrogen cyanate ranging from 0.007 %.

Most of the starch is used in production alcoholic drinks, or in ethanol production or it can serve as energy source. The ethanol produced in Brazil from Manioc is used in petrol mixture and it is called as gasohol which is used as motorcar fuel. Starch is an important substrate in production of amylases, ethanol and antibiotic synthesis which are produced by actinomycetes. This is because starch accumulates slowly and it doesn't create catabolite repression. The starch which is of lower quality is used in ethanol and biomass production. In order to be utilized by the microorganisms, starch has to be degraded into maltose and glucose either by enzymic hydrolysis or chemical hydrolysis.

Inulin

Inulin is colorless and it exists as microcrystalline substances. It is soluble in warm water. It is formed as a polysaccharide of stolon plantslike chicory, artichokes, corn, tuber, Taraxacum and Inula. It is poly β D fructan and is non-reducing. It is made up of 30 fructose sub units. Inulase helps in splitting of inulin to D fructose. Inulin can be obtained from artichoke which has tubers from Heliantus tuberosus. It is used for food stuff production and also as fodder. It acts as a carbon source in microbial growth. The plants and tubers constitute about 20% weight%. The main limitation of tubers is their spoilage in short time. Microorganisms which produce inulase can easily utilize inulin. Inulin is used in yeast cultivation and ethanol production.

Cellulose

The energy produced in the photosynthesis process is stored in the form of cellulose. The result of photosynthesis which is in excess even after respiration all over the Earth is 20×10^{10} t of lignocelluloses and half of this composition is cellulose. Cellulose is obtained from woody as well as non-woody plants and also from wastes from food industry, agriculture, wood industry and forestry. The wastes from the wood processing have 40% of mass of wood and in rare cases they yield higher cellulose content than wood. Pure form of cellulose cannot be achieved where as the complexes such as lignin and hemicelluloses are available. Plants constitute celllose, hemicellulose and lignin in 4:3:3 ratio. This ratio can vary based on the biomass of lignocellulose. Cellulose is a homopolymer and it is linear. The cellulose filaments are enclosed by hemicellulose and lignin. Lignin consists of benene rings or polyphenols which are highly oxidized. Hemicellulose constitutes sacchardie acids, pentoses and hexoses. Cellulose is hard to hydrolyze and hemicellulose is easy of hydrolyze. Lignin cannot be easily hydrolyzed. The hydrolysis of cellulose is difficult because of the following

- It is a complex structure and about 80% of the cellulose exists in crystalline form.
- The filaments are protected by lignin.

The prehydrolysis or pretreatment is the set of the procedures which are targeted to release cellulose from the enclosure of lignocellulose complexes. It also includes transforming the crystalline form into the amorphous state. Some of the procedures are given below

- Mechanical procedure includes cutting and grinding
- Physical procedure includes exposure to radiation, pressure and heat
- Physico chemical and mechanical procedures
- Biological procedure which includes enzymes such as hemicellulase and ligninases

Often the combinations of techniques are employed to yield complex effect.

The mechanical procedures involve grinding with a hammer in order to reduce particle size and it also helps in increasing the susceptibility of hydrolysis to occur. This process doesn't change the proportion of the component which is crystalline. Colloid milling or wet milling results in particle diminution and it also increases the susceptibility of hydrolysis, but this process is costlier. Krupnova and Sharkov proposed material grinding at 240^{0}C. The heating process itself makes the compound ineffective.

The physical procedures include dry heat or cleavage and wet heat or extrusion. In this process, the filaments are ruptured and cellulose is released in the form of small chunks. The effect in total is based on the degree of expansion. In the process of extrusion, the saturated steam is allowed to pass or penetrate into the material with pressure for 20 min and later pressure is released. The water which is present inside the substance or the material gets evaporated. This process disintegrates material rapidly into tiny pieces which exhibits more porosity. This process is more economical. The complexes which constitute lignocelluloses can be depolymerized with the help of gamma radiation and in higher doses. The dose of gamma radiation can be greater than 10 Mrad. If the dose is less than 1Mrad, then there won't be any significant effect and if the dose is greater than 50Mrad the cellulose losses its integrity and starts dissolving in water. The degree at which cellulose depolymerizes depends on the radiation dose applied. Due to the radiation, there will be free radical formation and major portion of the lignocelluloses gets solubilized. Bhatt et al studied the effect of gamma pretreatment on forest lignocelluloses biodegradation with the help of Aspergillus niger.

The chemical procedures are common in the process of lignocellulose pretreatment. The important advantage using these chemicals is that the effect is rapid and it also results in good production. The pretreatment can be done by either acid hydrolysis or alkaline hydrolysis. The acid hydrolysis process for the pretreatment was first used in 1913 in the USA. Alkaline pretreatment is generally done with NaOH. This treatment is used in straw processing to enhance digestibility especially in the ruminants. The pretreatment is costlier but it is one of the effective methods. If the NaOH concentration is greater than 8% w/w then there will be intramicellar swelling of filaments which contains cellulose and later depolymerization. If NaOH concentration is in between 0.1 to 0.5 v/v then there will be esterification of uronic acids and saponification of acetates. These reactions cause swelling of the structure and thereby loosening. Zinc oxide when added to alkaline solution enhances swelling of cellulose. The alkaline treatment is proved to be effective when hrad wood is used. This treatment depends on the levels of lignin. Excess amounts of alkaline solutions are used and wash water too. This will decrease the dry weight by 25%.

Some of the other methods are cellulose extraction from lignocelluloses complex followed by hemicellulose removal and later lignin oxidation. Cellulose can be hydrolyzed to glucose if dissolved. Cadoxene can be used for dissolving glucose efficiently. It is a mixture of 20 to 30% of alkaline aqueous solution of rthylene diamine and cadmium hydroxide. It helps in dissolving cellulose and when water is added there will be reprecipitation of cellulose in fine tufts. It has the capacity to dissolve 10% cellulose without any pH adjustment. It causes minimum damage to the cellulose and is stable. With this method, the preparation is easier and can easily regenerate. This solution is not toxic for enzyme cellulose but the cadmium molecules can exhibit negative effects.

One of the other agents which help in solubilization of cellulose is CSMS. It is a mixture of 6.2% Sodium sulphite, 7.5% caustic soda, 6.6% ferric chloride and 17% sodium tartrate in aqueous solution. The toxicity percent is low when compared with cadoxene. The preparation can be made in solid form and hence transport along with handling and storage is easier. After the solubilization of cellulose, methanol can be added to precipitate cellulose. Hydrogen peroxide (H_2O_2), sodium hypochlorite (NaClO) and sulphur dioxide (SO_2) are some of the other agents which disrupts lignocelluloses. If sulphur dioxide increases then the lignocelluloses become sensitive towards hydrolysis. NaClO is used to release lignin in case of sawdust and straw. It is effective at 0.5 M concentration and pH should be maintained at 9 at normal room temperature. Ozone acts on lignin rather than cellulose.

The best method is to combine methods so as to get all the advantages of chemical, physical amd mechanical processes. The lignocelluloses materials are grinded and then heated via steam feed inpresence of high pressures. This process can be followed by chemical treatments and it might be with the help of liquid or gaseous agents. One of the efficient procedures is ammonia treatment with a liquid and gaseous phase and the temperature is maintained at 132.4 ^{0}C and the pressure is at 11.28 ^{0}C. The lignin molecules can be removed from cellulose with the help of butanol by using butanol-water mixture in the ratio of 1:1 and also hydrochloric acid (HCl) in the concentration of 0.3%. This treatment is done at 170 ^{0}C for 60 minutes. The products can be separated easily since lignin can be accumulated in the phase which contains butanol and hemicelluloses get accumulated in the phase which is aqueous. The cellulose molecules remain in the phase which is solid.

The enzymic process in the production of cellulose is economical because there is no requirement of high temperatures and pressures. The degradation of lignin using enzymes was in process for 80 years. Phlebia radiata and Phanerochaete chrysosporium produces lignases and these enzymes degrades lignin with high rates. The limitation with the lignin production is that their structure is synthesized arbitrarily and randomly. Ligninase has the capacity to catalyse even in absence of oxygen and the mode of action is similar to the peroxidases. Ligninases can abstract electrons from substrates and they exhibit higher stability. Because of these advantages with ligninases, these enzymes are used in the recovery of glucose and cellulose. The cellulose which is produced can be hydrolyzed by acid hydrolysis or through enzymes to form glucose. Glucose can be form in scarce amounts during the lignocelluloses processing via physical and chemical methods. The solvolysis of wood by using hydrogen fluoride (HF) is one of the latest methods to hydrolyze cellulose into glucose. Another latest process is the swelling of cellulose with calcium chloride at saturated aqueous solution and then hydrolysis with HCl (gas) at lower temperatures. The product obtained constitutes 45% w/w of glucose and calcium chloride is extracted via electro osmosis. The enzymic hydrolysis is much useful when *Trichoderma* species were used since they contain cellulases. *Trichoderma Konigii, T. viride* and *T. reesei* are important species which possess cellulase system. The enzymes are

Carboxymethylcellulase or endo-β-1.4-glucanase or Cx cellulase

- Cellobiohydrolase or C1 cellulase or avicellase
- Cellobiase or β-1,4-glucosidase

The C1 cellulase is found in greater amounts in *Trichoderma*. It constitutes about 80% of the entire cellulose proteins. The hydrolysis of cellulose starts with splitting of β-1,4 bonds in presence of endoglucanase. The exoglucanase isolates cellobiose from the ends of the chain which are non-reducing. Exoglucanase and endoglucanase combine to produce cellobiose and oligosaccharides. Glucosidase hydrolyses cellobiose and other available oligomers to yield glucose. The enzyme synthesis possesses some regulatory mechanisms such as catabolite repression and induction and it is best to isolate certain strains which do not contain regulatory mechanisms. The main advantage of the chemical hydrolysis of cellulose is it speed and the whole process takes just 20 minutes whereas the enzyme mediated hydrolysis does require several hours. The capital cost which is required in chemical hydrolysis is higher and it has certain technical problems such as corrosion. The enzyme mediated hydrolysis requires production volumes of high intensity and the cost of equipment is cheap. Enzyme mediated hydrolysis is free from toxic substances where as the chemical hydrolysis result in toxic chemicals which are toxic for the microbial processes. Pure glucose may not be produced in case of enzyme hydrolysis since the starting material exhibits higher complexity.

Plant oils

The plant oils can be classified into slightly congealable or non-congealable, congealable and semicongealable. Among all the three semicongealable oils are most common. These oils contain lower content of unsaturated fatty acids. Linoleic acid and oleic acid are the examples of these oils. These oils are mostly obtained from soya beans, sunflower, and rape seed. The non-congealable oils are made up of 20% saturated fatty acids. The saturated oils are linoleic acid and oleic acid. Linoleic acid is presnt in minut amounts. These oils are abundant in peanuts. Congealable oils contain less amounts of saturated fatty acids and linolienic acid, linoleic acid are some of the unsaturated acids. The oleic acid content is much lower than the linolenic acid. It is found in poppy seed and hemp. The plant oils constitute mucous substances, chlorophyll and proteins.

These oils which are obtained from the plants are mainly used as carbon source in the production of antibiotics. The concentrations exhibited by these oils can be 10 to 30%. These oils can act as cosubstartes especially in the production of citric acid, beta carotene, cephalosporin, streptomycin, tetracycline, and penicillin. These oils if present in the nutrient media have the capability to protect the actinomycetes filaments against the mechanical damage especially by the shear forces. These oils can dissolve certain metabolites which exhibits inhibitory effects. Generally, the plant oils used in the nutrient media are palm oils and soya oils. The semicongealable oils pose certain problem during their transport especially in the piping process and so preheating is compulsory. The plant oils are also used as defoaming agents. These oils are consumed by the microorganisms and it is mandatory to replenish them in the nutrient media. Before framing certain formula, it is mandatory to see that the energy content of the oils to be used should be 2.5 folds higher as compared with the saccharides.

Nitrogen

The second most important ingredient present in the nutrient medium is the nitrogen. The can be produced from plants especially from the complicated substrates such as amino acids, peptides, polypeptides and proteins. Some of the substances which accompany nitrogen and present in the media which supports the microbial growth are trace elements, vitamins and lipids. The Nitrogen compounds can be etracted from tuber crops, oleaginous crops and cereals and flours. The nitrogen compounds are found in the bye products which are formed during the processing of raw materials such as starch and oils. Some of the raw materials are treated before adding them to the nutrient media such as corn steep liquor. Some of the specialized nitrogenous substances which occur naturally are yeast extract and soybean peptone. The cereal fluors constitutes about 10% of the protein component and they are concentrated in the aleurone layer and in the germ portion which contains low percent of polysaccharides and highest percent of vitamins, minerals and lipids. High percent of proteins are present in soybean flour which ranges from 44 to 47 % and it is the rich source of organic nitrogen. Soybean flour constitutes 32% of nitrogen free substances, ash substances, 6 % of ash substances and 1.2 % lipids. It also contains phosphatides, phosphorus, and calcium. Peanut meal is one of the materials which is of utmost importance and it is extracted post oil extraction and peanuts. It appears that these materials possess similar compounds but still they cannot be substituted in the microbial growth processes.

Soybean flour is an important ingredient in production of antibiotics no matter organisms like fungi or bacteria such as *Streptomyces* or *Bacillus*. The peanut meal is being used in production of chlortetracycline and it is avoided in production of tetracycline. Flours and cereal meals from wheat and rye are rarely used for the production of antibiotics.

Corn steep liquor

It is extracted as a bye product in maize starch production. It constitutes biologically active substances in various proportions and this quantity and quality is based on the maize quality. Corn steep do not comprise any fixed composition. It is used in fermentation of penicillin and it is a staple component for antibiotics production and also secondary metabolites. It contains huge amounts of ash substances, trace elements, reducing substances and nitrogen compounds. The amino acids constitute around 80 percent of the nitrogen and this proportion can be increased by the process of hydrolysis. The ratio of the amino nitrogen to that of total nitrogen is an important parameter which needs to be considered. The proportion is linked to the intensity of the lactic acid fermentation in the process of corn steeping. The corn steep liquor is heated to 100^{0}C and later it is cooled. Grey white precipitate gets deposited in the nutrient media which is not advisable. It can be removed by certain procedures. One Kilogram of corn steep liquor is treated with 2.5 liters of tap water and pH is kept at neutral with the help of sodium hydroxide solution. The entire solution is heated for 10 minutes and later cooled down followed by filtration through kieselguhr. The precipitate which is deposited on the filter is thoroughly washed to make up the solution to 5 liters. 5 liters of filtrate is equal to 1 kilogram of corn steep liquor. The precipitate comprises of magnesium phosphates and calcium phosphates. Post treatment, the precipitate is suitable for the cultivation of microorganisms and it is good for the biotransformation process and isolation of products. This treatment increases the production of antibiotics by three times.

Yeast extract

Yeast extract is produced by plasmolysis of brewer's yeast or backer's yeast. Yeast needs to be processed and the bitter substances should be removed with water at the neutral pH. The process of autolysis occurs in tanks with rigorous stirring. The temperature during this process is increased to 50 ^{0}C and kept at the same temperature for many hours. The temperature is again increased to 75 ^{0}C to inactive enzymes present in the yeast. Plasmolysis generally occurs when sodium chloride is present but the extract itself contains about 40 % of sodium chloride. The cell walls are isolated from the solution by either the process of centrifugation or filtration. The liquid is subjected to concentration at 3 7^{0}C to yield syrup and this prevents contamination and also protects heat sensitive substances. The yeast extract in the powder form is produced by drying with the help of spray drier or in vacuo. Due to its high cost, the yeast extract is used mostly on the laboratory scale and rarely in industrial scale.

Raw materials from animals

These raw materials are scarce than raw materials obtained from plants. These materials are generally represented by the secondary products obtained from food and agriculture industry. The main saccharide which is generated from these raw materials is lactose. Other raw materials include peptone, meat extract which serve as nitrogen source and animal fats. The waste materials which are of animal origin are produced while processing fish foods, poultry and meat. Since these raw materials have environmental properties which are unpleasant they are simply disposed from the stable and animal houses. These wastes generate methane when subjected to anaerobic conditions and it has the capacity to yield energy. The solid wastes which remain post processing can be used as manure but after sterilization. This waste can serve as substrate for cultivation of fungi on the surface. The liquid manure is being tested to replace the nutrient medium especially in production of ethanol.

Whey and Lactose

Lactose is present in milk of several mammals in the concentrations ranging from 3 % to 8 %. Both old and modern methods are used to produce whey. The curd is filtered before processing. Whey is generally heated to 45 ^{0}C and later centrifuged in order to extract fat. The concentration process is done with the help of vacuum evaporators and so there will be supersaturation which is much necessary for the crystallization of lactose. The whey which is concentrated at high temperature is allowed to be transferred into the crystallizers. In the crystallizers, it is agitated and allowed to cool in order to produce lactose crystals. After the procedure, the crystals are separated and later dried followed by grinding. The yield of lactose is 95% if produced via this procedure. The modern approach comprises of the same process but the proteins are removed by the process of reverse osmosis. Lactose is generally assimilated in prokaryotes such as microbe's especially lactic fermentation bacteria and Enterobacteriaceae members. *Kluyveromyces fragilis* is one of the yeasts which assimilate lactose. The lactose molecules serve as an important energy source in the penicillin biosynthesis.

Whey is produced as a byproduct in the process of casein production, curd and cheese making. The composition of whey varies and it depends on the method employed to produce cheese. Whey which is sweet is degraded into sour taste because of the conversion of lactose into lactic acid by the microbial action. Whey is dried before using it as an ingredient for nutrient media. Due to high water contantranging from 93 to 96 %

there is risk of contamination and also excess transportation costs. This is being used in the production of propionic acid, lactic acid and nisin which is an antibiotic and also in the nutrition materials for humans. It can be used as fodder especially in the small dairies.

Fats

The animal fats are less available than the plant oils and they are in solid states. Liquid fats are good for producing nutrient media and they are fish processing oil. Lower quality fish oil and lard oil. These fats are used as defoamers especially in the production of antibiotics.

Petrochemical and chemical

The production of proteins on an industrial scale from microorganisms has brought information to use the raw materials of petrochemical origins. These are available on a large scale and they do have stability. These substances are classified into precursors, trace elements, organic and inroganic sources of phosphorus and nitrogen, acids, alcohols, and gases. The gaseous substrates include hydrogen sulphide, hydrogen, carbon dioxide, carbon monoxide, methane. The methane gas is abundant in the oil fields and it is considered as natural gas. It is used as a preferential energy source. It can be produced by partial oxidation of crude oil, bituminous oils and coal at low temperatures. Carbon dioxide can act as carbon source in phototrophic bacteria which belongs to greenbacteria such as *Chlorobiineae* and purple bacteria such as *Rhodospirillineae*. These bacteria can be employed for production of hydrogen, sulphur and other pigments. Carbon monoxide is sued by aerobic bacteria such as *Pseudomonas* and *Bacillus*. Hydrogen sulphide can be used by *Chlorobium* bacteria. *Chlorobium thiosulphatophilum* has the capacity to assimilate carbon dioxide and hydrogen sulphide if autotrophic conditions persist. It can accumulate polyglucose and can produce sulphur. In absence of light, it produces three organic acids with attached acetic acid from polyglucose.

Alcohols and organic acids

Methanol can be produced by the oxidizing methane chemically. Methanol is being used in te production of polysaccharides, vitamins, organic acids such as alpha keto glutaric acid, fumaric acid and citric acid, amino acids, serine, threonine, glutamic acid and lysine. The large-scale use of methanol is not recommended due to its toxicity. Due to this, it is mandatory to add methanol in the form of doses.

Ethanol can be produced with the help of microorganisms which feed on starch containing raw materials. Ethanol is utilized in the production of cephalosporin, ergosterol, orotidine, orotic acid, polysaccharides, vitamins, tricarboxylic acids, amino acids such as glutamate and yeast used in human nutrition.

Acetic acid has numerous applications in the field of microbial production on an industrial scale. It can be obtained from *Clostridium thermoaceticum* and this is homoacetogenic microbe. *Candida lipolytica* is on one of the yeast members which are used to produce n-alkane such as amino acids, lipids, citric acid and other antibiotics.

Nitrogen sources (Inorganic and organic synthesis)

The microorganisms can grow in faster pace when grown in organic substrates rather than inorganic substrates. Ammonium sulphate, diammonium hydrogen phosphate, ammonia and urea are some of the nitrogen sources used in microbial production.

Trace elements

Some of the trace elements used in microbial production are iron, zinc, nickel, molybdenum, cobalt, chlorine, magnesium and phosphorus.

Control of raw materials

The quality of raw materials and their usability for microbial growth process is decided by biological test. The suitability of a batch of raw material can be done by checking cultivation trails which include three flasks filled with standard medium and the other three flasks are filled with batch of raw materials. The raw material suggested needs to attain 90 % identical properties to that of standard materials.

The raw materials can be stored in polythene bags and stored in dark upto 15 ^{0}C. The raw materials which are in liquid state should be stored underground silos, barrels, casks, and tanks.

REFERENCES

1. Bhatt, A. K., Bhalla, T. C., Agrawal, H. O., Sharma, N. (1992) Biotechnol. Techn. 6, 111.
2. Bla~ej, A., Ko~ik, M. (1985) Degradation reactions of cellulose and lignocellulose. In: Cellulose
3. and its Derivatives, Chichester.
4. Buzzini, P., Gobbetti, M., Rossi, J., Ribaldi, M. (1993) Biotechnol. Lett. 15, 151.
5. Chung, S. L., Myers, S. P. (1979) Dev. Ind. Microbiol. 20, 723.
6. Cuskey, S. M., Frein, E. M., Montenecourt, B. S., Eveleigh, D. E. (1982) Overproduction of cellulase
7. -- screening and selection. In: Overproduction of Microbial Products, London.
8. David, C., Fornasier, R., Greindl-Fallon, C., Vanlautem, N. (1985) Biotechnol. Bioeng. 27, 1591.
9. De Menezes, T. J. B. (1978) Process Biochem. 9, 24.
10. Dimmling, W. (1978) Starch/Stiirke 30, 401.
11. Enweta, C., Ayanru, D. K. G., Obokwe, C. O. (1992) Acta Biotechnol. 12, 127.
12. Garves, K., Dietrichs, H. H. (1975) Holzforschung 29, 39.
13. Goto, T., Harada, H., Saiki, H. (1974) Bull. Kyoto Univ. Forrest. 46, 153.
14. Gould, M. J. (1985) Biotechnol. Bioeng. 27, 225.
15. Grous, W. R., Converse, A. O., Grethlein, H. E. (1986) Enzyme Microb. Technol. 8, 274.
16. Han, Y. W. (1978) Adv. Appl. Microbiol. 23, 119.
17. Hardt, H., Lamport, D. T. A. (1982) Biotechnol. Bioeng. 24, 903.
18. Jayne, G., Neuschaffer, K. (1957) Die Makromol. Chemie 28, 71.
19. Kirk, T. K., Croan, S., Tien, M. (1986) Enzyme Microb. Technol. 8, 27.
20. Krupnova, A. V., Sharkov, V. I. (1964) Gidroliz. i lesokhim, prom. SSSR 17, 3.
21. Kurachi, M. (1959) Bull. Inst. Chem. Res. Kyoto Univ. 36, 163.

22. Lamptey, J., Moo-Young, M., Robinson, C. W. (1986) Biotechnol. Lett. 8, 525.
23. Leclerq, E. (1985) Process Biochem. 20 (3), 75.
24. Lee, K. J., Skotnicki, M. L., Tribe, D. E., Rogers, D. L. (1981) Biotechnol. Lett. 3, 207.
25. Maddox, I. S. (1980) Biotechnol. Lett. 2, 443.
26. Meyer, J. M., Abdallah, M. A. (1978) J. Gen. Microbiol. 107, 319.
27. Miescher, G. M. (1975) U. S. Patent 3 929 575.
28. Milgrom, L. (1985) New Scientist 1, 16.
29. Millet, M. A., Backer, A. J., Satter, L. D. (1975) Bioteehnol. Bioeng. Symp. 5, 193.
30. Pfeffer, J. T., Liebman, J. C. (1976) Resources Recov. Conserv. 1, 295.
31. Prescott, S., Dunn, C. (1959) Industrial Microbiology, New York.
32. Puri, V. P. (1983) Biotechnol. Lett. 5, 773.
33. Ratledge, C. (1977) Ann. Rep. Ferm. Proc. 1, 49.
34. Reese, E. T., Mandels, M., Weiss, A. H. (1978) Adv. Biochem. Eng. 8, 181.
35. Rogalski, J., Hattaka, A., Wojtas-Wasilevska, M., Leonovicz, A. (1993) Acta Biotechnol. 13, 41.
36. Rotgers, P. B., Knowles, J. C. (1978) J. Gen. Microbiol. 108, 261.
37. Saddler, J. N. (1986) Microbiol. Sci. 3, 84.
38. Schlegel, H. G. (1992) Allgemeine Mikrobiologie, Stuttgart.
39. Shennan, L., Levi, J. D. (1974) Progr. Ind. Microbiol. 13, 3.
40. Sikyta, B. (1983) Methods in Industrial Microbiology, Chichester.
41. Toran Diaz, I., Jain, V. K., Allavis, J.-J., Baratti, J. (1985) Biotechnol. Lett. 7, 527.
42. Tsao, G. T., Ladisch, M., Ladisch, K., Hsu, T. A., Dale, B., Chou, T. (1978) Ann. Rev. Ferm. Proc.
43. Uchio, R. (1976) Japanese Patent 7 638 488.
44. Underkofler, L. A., Hickey, R. J. (1954) Industrial Fermentation 2, New York.

45. Voget, C. E., Mignone, C. F., Ertola, R. J. (1985) Biotechnol. Lett. 7, 43.
46. Wayman, M., Parekh, S., Chornat, E., Overend, R.P. (1986) Biotechnol. Lett. 8, 749.
47. Weimer, P. J., Chou, Y. (1986) Appl. Environ. Microbiol. 52, 733.
48. Angelov, A. I., Karadjov, G. I. & Roshkova, Z. G. (1996) Strain selection of baker's yeast with improved technological properties. *Food Research International* **29**, 235–239.
49. Batt, C. A. & Sinskey, A. J. (1984) Use of biotechnology in the production of single cell protein. Food Technology 38, 108–111.
50. Mizuno, T. (1995) Shiitake, Lentinus edodes: functional properties for medicinal and food purposes. Food Review International 11, 111–128.
51. Royse, D. J. (1995) Specialty mushrooms: cultivation on synthetic substrate in the USA and Japan. Interdiscipline Science Reviews 20, 1–10.
52. Solomons, G. L. (1983) Single cell protein. CRC Critical Reviews in Biotechnology 1, 21–58.
53. Trinci, P. J. (1991) Quorn mycoprotein. Mycologist 5, 106–109.
54. Chang, S. T. & Hayes, W. A. (ed.) (1978) The Biology and Cultivation of Edible Mushrooms. Academic Press, New York.
55. Chang, S. T. & Miles, P. G. (1989) Edible Mushrooms and their Cultivation. CRC Press, Boca Raton, FL.
56. Crueger, W. & Crueger, A. (1989) Biotechnology: A Textbook of Industrial Microbiology, 2nd edition. Sinauer, Sunderland, MA.
57. Demain, A. L., Davies, J. E. & Atlas, R. M. (1999) Manual of Industrial Microbiology and Biotechnology. American Society for Microbiology, Washington, DC. Large, P. J. & Bamforth, C. W. (1988) Methylotrophy and Biotechnology. Longman Scientific and Technical, London.
58. Vonshak, A. (ed.) (1997) Spirulina platensis (Arthrospira): Physiology, Cell Biology and Biotechnology. Taylor & Francis, London.
59. Wainwright, M. (1992) An Introduction to Fungal Biotechnology. Wiley, Chichester.

CHAPTER 6

DEVICES FOR PRODUCT FORMATION

INTRODUCTION

The culture devices are used to obtain a desired biological process which is much complicated. These devices help in deposition of microbial biomass, animal cells and plant cells along with their respective products. The selection of the culture device is very important and it should be based on technological factors, chemical, physical and biological factors. The types of cells which can be cultures are microorganisms, plant and animal cells. The microorganisms include bacteria and fungi. The substrates which can be used are solid, semi liquid, liquid and gaseous in nature. The oxygen supply is an important factor since the culture can be of aerobic and anaerobic. The aerobic cultures can be classified into surface and submerged. The demand of aseptic conditions is a crucial factor which determines the selection of cultivation devices. They can be operated in complete aseptic conditions or production-aseptic or partially aseptic or completely non-sterile. The scale at which the devices are to be used is crucial. They can be used in laboratory scale, pilot plant scale and production scale. The choice of technological configuration is also important and it can classify into batch, fed batch, semicontinuous and continuous. The choice of the device takes into consideration the size as well as the morphology of the cells. It means that the cells to be grown can be either unicellular such as bacteria and yeasts or filamentous such as actinomycetes or it can be multi cellular filaments such as higher fungi. The choice of cultivation devices needs to consider the structure of mycelium and if the mycelium is prone to mechanical stress and damage while culture growth.

An important factor is that the type of end product to be produced. The end product can be simply a cluster of biomass or it can be secreted into the medium. The cultivation devices can be classified as per the form and type of enzymes or the biocatalyst used. The bioreactors may help in culturing of free cells or the bioreactors can have enzymes which are immobilized or simply cells. Another deciding factor is the type of substrate to be used and it can be of solid nature, liquid or gaseous. If solid or liquid or semi liquid ingredients are used then there will be negligible problems. The problems are presented often if gaseous substrates such as methane, carbon dioxide, oxygen or air are used. Some microorganisms yield biomass and related products when grown on solid substrates and they cannot produce or grow when kept submerged. The devices which support the culture growth on the surface of the liquid and solid substrate is cheaper than the devices which support the microbial growth in submerged conditions and the growth is relatively slow in culture devices which support submerged conditions.

On of the important feature is that the cells relation with the oxygen. The oxygen supply should be maintained strictly if the microbe grows in aerobic conditions. Oxygen supply is a costly substrate so it is essential to take necessary steps while selecting the cultivation devices. The devices which provide surface aeration can be used in cases often isolated. Some of the microorganisms such as microaerophiles may not require oxygen but they can grow and can yield metabolites if oxygen is present. Oxygen is not required for production of anaerobic microorganisms and hence these microorganisms can be flushed from the cultivation device with the help of inert gas. The complexity involved in the design of culture devices depends on the needs of aseptic conditions to be maintained. In case of culture of pathological organisms in the culture devices it is mandatory to see that none of the pathologically active microbe gets into the environment. These devices are complicated in terms of pressure stabilization on medium, pressure of air and steam inside the device, foam removal and blockage of valves and joints. Some of the cultivation devices are simpler which maintain aseptic conditions. The sterilization of the nutrient media can be achieved by boiling and the air is also sterilized. The conditions of sterilization are maintained strictly while producing biomass under conditions which are ambient such as limited substrates, extreme temperatures and low or high pH which stops the growth of harmful microflora. In some processes such as waste water treatment normal micro flora which is available in natural conditions do exist and so the complete sterile conditions are not possible. The elaboration of the microbe growth process can be scaled up with the help of the cultivation devices. The increase of volume will be a gradual process and it is generally increased by a factor of 10.

The reactors used in the laboratory will be in the volumes of 50l, 500ml and the pilot scale bioreactors will range between 50l and 5 cubic meters whereas the production bioreactors will range from 5 cubic meters to 3000 cubic meters. If the vaccine is to be produced then the production bioreactor is about 100l. In case of microbial protein production, the vessel range is around 20 cubic meters. If the aseptic conditions complicate the devices in which the batch culture takes place then one can forecast more complexity in semi continuous and continuous cultures. One of the main factors which discourage use of such a type of cultures is the cost of the devices which are to be used for cultivation of microorganisms on a continuous scale. Because of the complexity inherited through the microbial process it is advised via the experience gained to use single stage continuous device for the biomass production and the end products. Sometimes multistage configurations are needed for the product formation.

Devices with solid substrates

The microbial cultivation on the surface which is solidified by silica gel, gelatin and agar are generally used for identification, classification of the microorganisms. The solid substrates are used for vaccine production by growing pathological organisms. Some of the bulk materials which are used for fungi cultivation are cellulosic wastes, straw, and bran. Solid substrates are used in coal desulphuration, microbial ore leaching and composting and also solid agricultural waste processing. The cultivation of the microorganisms on the solid substrate is done by using trays, cylinders, and flasks under partially aseptic or non-sterile conditions or under complete aseptic conditions. The microbial cultivation under aseptic conditions requires pre-sterilization of the device and the substrate along with the air and sterilization during sampling. The devices which are used for stationary culture especially on the solid substrate are simpler. The cultivation of the microorganisms is done with the help of metal plates which have the area of dm^2 or some square meters and are placed one above the other and it is arranged in a thermostated chamber. The thickness of the substrate layer which is wet will range around 4 to 8 cm. The moisture and temperature are regulated constantly with the help of moist air which is theromostated. The solid substrate is sterilized with the help of autoclave and the plates are also sterilized with the same process and then inoculation is done with the production culture.

If aseptic conditions are not needed then wooden trays can be used instead of metal trays. These methods are used to produce champignons which are edible fungi. The spaces which are formed in the wooden plates or the wire nets are filled with shredded straw which is a moistened surface. Inoculum is added till the space is filled. The microbial cultivation is stopped when the culture is spread on the media without leaving any spaces. Microbial ore leaching is one of the stationary cultivation techniques on the solid substrate. The dumb heads are formed after the mining process. These dumb heads comprise of metals in low concentrations and they are inoculated with the microorganisms which leach the metal in the form of their sulphates.

Devices with solid substrate mixtures

The cultivation device which belongs to this category is the vertical mixed bioreactor. It is a vessel with bottom which is perforated and the interior space is filled with solid substrate which is moistened. Air is supplied via perforated lower bottom and the content is effectively stirred with the help of motor or forks. The stirring is done with low intensity in order to avoid breakage of filaments. Stirring also helps in creating crevices in the surface of the solid substrate and hence allowing air to pass and the carbon dioxide formed is removed and also the culture overheating is prevented. In case of horizontally mixed bioreactor, the mixing is done by baffles and low revolution stirrer. The air is allowed to pass over the surface of the substrate preferably at one end the carbon dioxide produced is allowed to leave from the other end. Because of lower efficiency, these types of bioreactors are less often used. The tubular bioreactor is efficient and uses the entire work space and it helps in continuous cultures. A tube is fitted with a thread and it mixes culture and hence allowing the microbe to grow all over the space. The substrate is passed via tube at the same time. This bioreactor contains inlet and outlet for allowing steam and it condenses out while sterilization, a void or the opening is present for the supply of substrate and an outlet lets the processed substrate out of the bioreactor.

Surface culture using liquid substrates

The surface cultivation on the liquid media has gained importance in the recent years. This type of microbial cultivation is done in the bioreactors which has thin liquid layer, bags, bowls, pans, bottles, and flasks. The bioreactors which are used for this purpose are blade bioreactors, film, and disc bioreactors. These bioreactors help in ensuring effective movement of surface of the liquid culture. The initial costs for construction of these devices are cheap and the operation costs are less. The only limitation is that it possesses low velocity of production and growth. High concentrations of biomass per unit medium used are achieved in this process.

In a laboratory scale, the cultivations are done in Roux flasks, Fernbach flask and Erlenmeyer flasks since these flasks offer higher liquid surface area. The microbial growth depends on the intensity of gas exchange, surface area and the medium layer. The cultivation can be done in larger pans which are manufactured using pure grade aluminum and the walls are coated with zinc coated iron sheets which prevent corrosion and also the insertion of metal ions from the container can be averted. The surface area of each pan is 10 square meters and the height is about 20 cm. The culture liquid layer is maintained at 7 to 12 cm. Prior to use the pans are sterilized chemically or by using steam. The medium can be sterilized using central sterilizer. The sterile medium can be piped into pans which are arranged in the chambers. These chambers are aerated by sterile air and the relative humidity of 60 % is maintained at a constant temperature. The air supply is maintained at a rate of 15 cubic meter/ square meters per hour based on the process selected. This method can also be used for production of citric acid. Schroeder along with Brandl has designed a type of culture system which helps contamination free processes. The system proposed by them has pans which are staked one above the other inside a closed metal tank. The sterilized medium along with the inoculation is allowed to pass onto the pans from the top until the pans are completely filled. The tank is allowed to flush with the help of sterile air. The system is complex technically and the applications are limited. The growth of microorganisms on liquid media as stationary surface culture was explained by Kybel. The microbial culture is carried out with the help of plastic bags which are sealed. These bags can be used for industrial and laboratory applications. This device consists of air flow meter, regulating valve, air filter, cultivation bag, bypass valve, duplicator, manipulation bench, storage transporter and measuring valve. The transporter helps in positioning of bags. These bags are transported into the incubation space and later into the processing unit. The aeration is regulated by needle valve and cooling is done by cooling coils. The advantages of this process are

- The contamination from the external environment and from the process is low
- The biomechanical process can be regulated
- There is no need of sterilizing bags since they are already sterilized
- It is more than beneficial when used for fungal spore production

Bioreactors with thin layers

The devices which such as blade bioreactors, disc and rotating cylinder are the examples of this type. This type of devices is used when nutrient media is highly viscous in nature and when the microbial cultures are sensitive to shear stress. The rotating cylinder type of bioreactors is explained by Gorbach, Spencer, Sallans and Phillips. They are tube reactors and are enclosed with heat exchange jackets. They are fitted via a rotating cylinder and this cylinder dips into the medium where the microorganisms grow on the surface. The bioreactor contains outer tube and it will be revolving. It also contains inner cylinder which is stationary. The microorganisms are allowed to grow on the cylinder's surface as a thin layer and the renewal of nutrients is done due to continuous dipping in the nutrient medium. The revolving speed of the cylinder is around 7 Hz and the speed is low in the initial stages so that the microorganisms won't be washed off from the surface of the cylinder. The growing layer thickness is generally around 2 mm. If the layers are thick, there will be problem with the oxygen supply. The main advantage of this type of bioreactor is that very less foam is formed. The carbon dioxide which is produced in this process does not dissolve in the liquid and hence the supply of oxygen to the microorganisms is good. The main problem occurs during the collection of samples of biomass from the surface of the cylinder because of the rotation. These bioreactors are not yet used in the industrial scale.

The disc bioreactor contains revolving discs which are linked with a shaft. The microorganisms grow on the surface of the revolving discs which revolve slowly and these discs dip in the medium constantly. It contains a glass tube which is T shaped and this tube is sealed in both ends. These ends are sealed with the help of stainless steel plates. The revolving shaft is arranged in the center of the steel plates. The revolving shaft is driven by an engine and the revolution velocity is adjusted as per the need. The stainless steel or plastic discs are attached to the shaft. The discs which are in the middle portion are classified into four fragments or segments. While sampling one of the segments is taken out with the help of sampling bag. This segment is replaced using another segment which is sterilized. Each disc on the shaft is divided by rings and screws are used for fixing end plates. The liquid surface is arranged below the shaft. The sterile air is supplied through an opening and inoculum is also supplied through an opening above the surface of the nutrient medium. In the other end, the vessel contains an outlet which allows air to pass. The sampling port is present in the lower part. These bioreactors are being used for treating waste water and this system is to be tested for the production of metabolites. The principle behind the blade bioreactor follows the same principle used in disc bioreactors. The microorganisms grow on the blade's surface. The impeller revolves slowly and is immersed in the culture medium and it is semi immersed into the nutrient medium. The transfer of oxygen is good because of the fact that they grow as thin layers. The main applications of this type of bioreactors are almost same as the disc reactors. The culture devices which have a thin liquid layer can also include flat surfaces which are tilted. The culture medium slowly trickles along the line. This system is used for cultivation of autotrophic microorganisms via open field cultivation. Submerged cultivation is widely used method for the cultivation of microbial biomass and the metabolites. Submerged cultivation produces high rates of microbial biomass and their products. The process of stirring or air feed helps to transfer oxygen at an satisfactory rate and also it helps in homogenization of the contents of bioreactor. The submerged cultures can be performed using various flasks which are placed on the shakers.

Flasks and shakers

Perquin and Kluyver were the first scientists to use the flasks for microbial cultivation on the shakers. This method holds good for both animal cells and plant cells. The cultivation in flask can occur in large series. This help in evaluation and screening of microorganisms, inoculum, raw materials, and media composition. The microbial cultivation can be performed on shakers and can be conducted in bottles, flasks, and test tubes. In general culture flasks and Erlenmeyer flasks with volumes ranging from 250 to 500 m can be used. The oxygen transfer rate in the flasks can be governed by mechanical characteristics, shape of the flask and the ratio of total volume to the medium. The oxygen transfer rate is high if the flasks has flat bottom and it is lower if the bottom is round. The rate can be increased by arranging baffles in different types of flasks. It should also be observed that these baffles results in foaming and this will be detrimental to the oxygen transfer.

Glass flasks are being used widely and they are cheaper to use. These flasks are transparent and hence providing a chance to observe the processes such as growth of microorganisms, mixing of media, sterilization followed by washing. The glass flask needs to withstand heat, acid, alkali and should not release trace elements. Cotton plugs are used to close the flasks and those forms an oxygen barrier since oxygen needs to get diffused into the plug prior to culture. The wetting of cotton plugs with the nutrient medium should be avoided. Some of the plug materials other than cotton are synthetic fiber materials, gauze, polyurethane foam, and glass wool. The culture which is present in flasks loss of water by the process of evaporation. This evaporation causes decrease of culture fluid and hence altering ratio of total flask volume to the medium. This alteration will change the oxygen transfer rate and hence microbial growth. Adding additional water will help to compensate this water loss.

The shakers should be reliable, quiet and fault free operation devoid of vibrations even ran for longer periods of time. Rockers are the simplest type of shakers. They are used when microorganisms are to be cultivated in test tubes. These rockers provide pendulum like motion. The oxygen transfer rate is low when compared with other shakers. The shakers can be classified into rotary shaker and reciprocal shakers. The reciprocal shakers have mechanical simplicity and it is cheaper. The stroke frequency and the amplitude are the important parameters. When the amplitude and stroke frequency are increased, the rate of oxygen transfer also rises to threshold level. The amplitude is maintained at 30 to 80 mm and the stroke frequency is maintained at 1 to 2 Hz. The reciprocal shakers are used mainly to grow yeasts and bacteria which are unicellular organisms. While fungi cultivation via flask cultivation the walls might be covered with the culture and seldom forms solid pellicle on the surface of the medium. The rotary shakers provide enhanced rates of transfer of oxygen and these shakers are used widely. The eccentricity and revolution frequency are the key parameters to be considered in while using rotary shakers. The revolution frequency is to be maintained at 1.5 to 7 Hz and the eccentricity needs to be managed at 10 to 15 mm. The rotary and reciprocal shakers are attached with revolution frequency meters or the stroke frequency meters. These meters have automatic cut-off timers, programmable revolution frequency and regulation of stroke.

The culture flasks are generally placed on the trays and different types of brackets as per the size and type of the flasks. Tilting of flasks helps in enhanced oxygen transfer and the tilting angle can be from 15 to 30^0 C. The trays are generally fixed on the vertical plates. The trays which are used in the reciprocal shakers are rectangular and the shape of rotary shakers is rectangular or circular. The plates on each shaker can be about 1 to 4 in number. The use of multiple shakers can save space whereas the use of single flask can help the experimenter to manipulate the flask in different ways if so desire. The single plate shakers can be helpful and handy if we need to add accessories such as nutrient stock solution supply, air feed, gas feed and pH. The temperature is maintained constantly by installing shakers in the rooms which are thermostated. Smaller flasks are generally fitted with constant temperature cabinets or with water baths. Since air and glass are not good conductors of heat, abundant air circulation needs to be maintained in the room to avoid uniform spread of heat and therefore uniform temperature can be maintained. The flask which is placed in the center of the multi plate shaker can exhibit higher temperatures often ranging from 2 to 3 degrees than other flasks and this is due to maximum growth of culture.

Large volume cultivation devices

These devices are explained by Sikyta and Kybal. This device is suitable for cultivating cells which are sensitive to the mechanical stress. The plastic bags are filled with the nutrient medium which is sterilized previously and then inoculated. They are kept on a plate which has regular swing frequency inside a room which has constant temperature. The swinging motions make sure efficient mixing of cultures and also help in enhanced transfer of oxygen. The contamination is reduced because of less manipulation and configuration. The device consists of three parts. They are control panel, box, and the stand. The control panel is constructed into the chamber lid. The front part generally contains control elements which are used for heating and regulation of the swing. In the back side, outlet is present and it is equipped with two flaps which are important to regulate the intensity of the aeration. The front side of the box is fitted with a door which is transparent.

Bioreactors equipped with mechanical mixing

These bioreactors are used in both research and production purposes. The ingredients of the bioreactor are mixed with different types of impellers which have different shapes and placed at different places on the central shaft. The bioreactors equipped with mechanical mixing can be classified as per the size such as production types, pilot plant and laboratory.

The laboratory bioreactors are good suit for research purposes. These serve as the intermediate steps during the process of scale-up. They demand low operation and investment costs. These factors constitute to give several advantages to use these types of reactors. The working volume provided by these types of reactors ranges from 1 to 20 liters. There are two types of reactors namely portable reactors and stationary reactors. The stationary bioreactors are arranged on the stands and they are fitted with coils or the jackets for the sterilization and to regulate temperature. The portable bioreactors are installed in the form of batteries and they are autoclaved with or without medium. After the autoclaving process, the arranged place in the water bath which is maintained at constant temperature.

The primary requirements which are to be followed by nay bioreactor is

- The interior equipment, geometrical proportions, shape should mimic the industrial scale bioreactors.
- They should provide flexibility so that many varieties of proposed factors can easily be studied.
- They should be small in size so that it can be economically feasible
- They should be simple to clean, dismantle and assemble
- They should maintain a large working volume so that the sample procedure can be done effectively.
- The temperature regulation, foam breaking and pH are maintained by the controlling devices.
- They should be contamination free.

The main components of the laboratory bioreactor are the inoculation ports, sampling, baffles, air inlet, motor, seal chamber, impeller, lid and vessel. The vessel is made using borosilicate glass which is of high quality. In case of large vessels, stainless steel is used and in some cases both steel and glass are used. The bottom of the vessel is concave and flange is fitted on the upper rim. The cylinders contain both upper and lower lids which are metallic. If the vessel need to be maintained at overpressure, then it is essential that the vessel is to be encased with a protective basket. The lid is flat in shape and it is fitted to the vessel with the help of screws. The packing is placed in the space between the vessel and the lid and the packing is made up of Neoprene or Teflon. In case of the cylindrical vessels, the upper lid and lower lid are attached with a retaining rod. The agitation can be done by either short shaft in the center or with the help of impeller. The impellers present on the top are easier to install. The impeller shaft needs to be sealed within a packing chamber with numerous configurations. The process of electromagnetic coupling can be employed to run the impeller shaft. The aforesaid arrangement is advantageous because there is less chance of contamination if there are any leaks in shaft.

The laminar liquid flow of the fluid is reduced by using four baffles or it can be one. These baffles can be fitted into the inner layer of the lid or vessel and can be fitted by the screws. They can be made as a part of the vessel. The air is allowed inside either from the lower lid or the upper lid. The tube can be made as a ring-shaped structure. Prior to the air inflow, the air is filtered using plate filters or sintered glass or cotton wool. The overpressure of air is regulated by maintain pressure valve. The culture samples can be collected by sampling tube from the bottom lid or via the upper lid. The first fraction of the sample is not used.

Constant temperature can be maintained by various methods in the bioreactor. The temperatures can be regulated by the flow of warm water and cold water or by using a electric heater and also cooling equipment.

The foam can be broken down by using mechanical methods such as disc shaped foam breaker or the blade shaped foam breakers which are made up of wires and other economically feasible materials. There are fitted on the impeller shaft and so that the foam breakers revolve at the same speed as the impeller.

The impeller which is attached to the shaft is run by an electric motor. The impeller and the motor can be coupled directly or in some cases the shaft is assisted with the belt pulley. One of the best defoaming systems is Turbosep and it is devised by Dominick Hunter Filters Ltd. It can be used with the bioreactors which offer a flow rate of 4000 liter per minute. The deforming device can enhance the working volume to 30%.

Pilot plant

The pilot plant bioreactors can help to modify, complete, and verify the data so that the scale up process can be forwarded. These bioreactors possess higher speeds offered by the impeller so as to mimic the conditions which do exist in the large-scale reactors. Various types of agitators' baffle dimensions, aeration mechanics and others can be studied in detail in the pilot plant bioreactors. The key feature of any microbial synthesis is that there should be no foreign microbe while cultivation. The large bioreactors are sensitive to contamination due to the complexity. The reliability can be checked by sterility test. The bioreactors are supplied with rich medium and it is without inoculation. The processes such as sampling, air feed and agitation is carried out as the same process which occurs in the large-scale bioreactors. The samples from the bioreactor are subjected to the sterility test.

The fitting and pipelines are sterilized along with the air filters. Post medium sterilization, the outlet valve which is present at the bottom is opened so the air bubbles are removed and the medium can be cooled. If the pressure inside the bioreactor lowers to 0.05 MPa then the air is pumped using the aeration devices. In the process of medium sterilization, the connecting lines and the valves are flushed with the steam. Flange packings are replaced with welding. The packing's should not be made with the porous material. They need to have high quality and should be resistant to higher temperatures which is mandatory in the process of sterilization. They should not crack or loose elasticity during the process of sterilization. The valves are mounted under the steam seal and it is horizontal position. The safety valves, manometers and back pressure regulators are fitted into the tubes.

Industrial bioreactors

The agitation process in the large scale or the production scale bioreactors are based on the various types of baffles and filters used. The six bladed and open types with diameter of one half the diameter of the vessel are widely used. The radial flow is achieved by the forces created by the turbine blades. It results in the dispersing effect. It is not uncommon to use three or almost two turbines on a single shaft. The impellers are installed into a separate cylinder. Each impeller consists of a separate mixing field. This allows multistage gas dispersion. In some cases, the baffles can be removed and this helps in stirring of turbines and also eases the formation of vortex at the center. This will result in choking of impeller by air and may lead to inadvertent energy loss. Vogel Busch bioreactors are installed with conical impellers which are hollow whereas the Wald of vessels have hollow bent paddles. These systems provide a suction effect which is behind the impeller. Air is allowed to flow into the fluid with the help of hollow impeller. The path taken by bubbles is spiral and this helps in long hold up times of sir bubbles and hence increased utilization of oxygen. A cylindrical diffuser which is circulating is installed with the Wald of bioreactor and it acts as a foam breaker. If the impeller is installed at the lower part of the diffuser, higher rates of oxygen transfer is achieved. The impeller can be bottom driven or top driven.

The bioreactor which is simple and fitted with pneumatic mixing apparatus is the bubble column. The transfer of oxygen is low in case of mechanically agitated bioreactor. The columns are generally installed with static mixers and the sieve plates. This will increase the bubble hold up time. The principle of air-life for aeration and circulation was proposed by Seidel and Scholler. This principle is based on the difference between specific mass of fluid alone and the whole dispersion. Central air feed and peripheral air feed are the two types of arrangements. It is used in case of anaerobic waste water treatments. A bioreactor which ranges from 2000 cubic meters to 8000 cubic meters can suffice the waste water delivered from a city with about million populations. Recently the bioreactors which are equipped with hydrodynamic mixing have become common. These bioreactors are classified into two types; they are deep jet bioreactor and loop bioreactor. The deep jet bioreactor uses external pump for efficient fluid mixing. Higher rates of oxygen transfer can be achieved by using a narrow jet of the fluid along with the air and makes its way to the bioreactor.

In laboratory devices, the microorganisms are free from the oxygen. This property can be achieved by culturing in the flasks and test tubes which are filled with the inert gases such as nitrogen, hydrogen and carbon dioxide. The flasks or the test tubes are cultures on the shakers or just stationary states. The closure of flask should be hermetically sealed. In case of removal of gas from the test tubes, the mixture of gases is flushed out by using inert gases. Longin, Henriquest and Bhatnagar has devised a system for cultivating strict anaerobes. The design of the large devices which are used for the anaerobic microbial growth process has to take into account gaseous and liquid substances. The bioreactors which are used or production of organic acids and solvents are to be fitted with stainless steel vessels which have low speed agitators. These help in dispersion of microorganisms throughout the volume of the vessel and the medium. The vessels are fitted with sampling and inoculation assembly, temperature and pH sensors and gas outlets. The anaerobic processing of the wastes is done by tower bioreactors. The microorganisms are allowed to grow as a film. Different devices have different inlets for substrates.

REFERENCES

1. Adler, I., Fiechter, A. (1983) Swiss Biotechnol. 2 (2), 17.
2. Antonie, R. L. (1975) Fixed Biological Surface -- Waste Water Treatment, Boca Raton.
3. Bhatnagar, L., Henriquet, M., Longin, R. (1983) Biotechnol. Lett. 5, 39.
4. Blain, J. A., Anderson, J. G., Todd, J. R., Divers, M. (1979) Biotechnol. Lett. 1, 269.
5. Blakebrough, N. (1969) Design of laboratory fermentors. In: Methods in Microbiology 1, London.
6. Britz, T. J., Meyer, L. C., Botes, P. J. (1983) Biotechnol. Lett. 5, 113.
7. Comberbach, D. M., Bu'Lock, J. D. (1984) Biotechnol. Lett. 6, 129.
8. Delicher, H.-H., Kroetsch, P., Popp, K. H., Stickel, R. (1974) Chem. Ing. Technik 46, 337.
9. Deschamps, A. M., Henno, P., Pernelle, C., Caignault, L., Lebeault, J. M. (1979) Biotechnol. Lett. 1, 239.
10. Einsele, A. (1984) Swiss Biotechnol. 2(4), 25.
11. Gorbach, G. (1969), Mschr. f. Brauerei 22, 22.
12. Guiot, S. R., Van den Berg, L. (1984) Biotechnol. Lett. 6, 161.
13. Hines, D. A. (1978) The large scale pressure cycle fermentor configuration. In: Biotechnology, Dechema Monographs 82, 55.
14. Jagusch, L., Pfischel, S., Schieferdecker, K., Schuster, H., Meyer, F. (1972) Czechoslovak Patent
15. 149 588.
16. Kennedy, K. J., Van den Berg, L. (1982) Biotechnol. Lett. 4, 137.
17. Kluyver, A. J., Perquin, L. H. C. (1933) Biochem. Zeit. 266, 68.
18. Kvasni~ka, J. (1972) Czechoslovak Patent 148 540.
19. Kybal, J., Sikyta, B. (1985) Biotechnol. Lett. 7, 467.
20. Kybal, J., Vl~ek, V. (1976) Biotechnol. Bioeng. 18, 1713.
21. Manchee, R. (1979) Trends Biochem. Sci. April.

22. Means, C. W., Savage, G. M., Reusser, F., Koepsell, H. J. (1962) Biotechnol. Bioeng. 4, 5.
23. Moser, A. (1982) Biotechnol. Lett. 4, 281.

CHAPTER 7

MICROBIAL PROCESSES

INTRODUCTION

The yield and the scale up process from pilot plant scale to the production scale can be affected by regulation of microbial processes, control and measurement. The lab trails include regulation of biological, physical and chemical factors. The operation costs are high but these costs are necessary to achieve the anticipated product yield. In order to achieve good regulation of the microbial processes, it is mandatory to

- One should perform cultivations with good instrumentalized systems
- One should correlate the obtained results with the data known regarding metabolic regulations and cell metabolism
- One should utilize the gained knowledge to enhance the optimal regulation which includes feedback regulation of the culture medium, analysis and continuous monitoring.

The choice of the method used for measurement and regulation is based on different factors such as type of microorganism, medium composition, substrate, product, process type and production device scale. The type of microorganism and the nutrient media composition will affect the method to be used to analyze the cell matter concentration, proliferation rate, growth rate, substrate type, the amount of substrate to be added while culturing, product type and concentrations of the product.

The measurement devices and the process of regulation do occur in non-aseptic conditions. These devices are much different from the devices which do require strict aseptic conditions. The process of measuring and the regulation of the biological factors, chemical and physical factors in case of small units differ much from the production scale or the large-scale devices. Automatic analyzers are not recommended in case of small devices since more culture volume is taken out during the process of sampling. Due to this limitation, these devices are installed with certain electrodes *in situ* in the culture. The large-scale bioreactors are installed with sensors at various points in the bioreactor. The sensors can be categorized as per

- The sensor location in the bioreactor or outside the bioreactor or on the bioreactor
- Data acquisition; continuous or discontinuous
- Ambient factors such as molecular biological, biological, chemical or physical
- Date type; primary or direct and secondary or derived data

Devices and Methods

The regulation and the measurement of the environmental factors are done with the help of the sensors and they are used in several industrial processes. The installation of sensors is a must to eliminate any foreign material or contaminant in the culture is possible. The chemical factors are measured using nuclear magnetic resonance, gas chromatography, spectrometric method or spectrophotometric method, calorimetry, catalytic sensors, semiconductor gas sensors, enzyme thermistors, enzyme electrodes, ion selective electrodes and automatic analyzers. Some of the methods are specific to the microbial processes which include both molecular biology techniques and biological techniques which help in measuring the cell composition and its respective components, populations and homogeneity, proliferation and so on. There are some methods which allows to measure collected samples and others deal with the continuous measurements on-line.

Automatic analyzers

The automatic analyzers can be of multiple purposes or for single purpose. The single purpose automatic analyzer is used to measure the elemental composition of the cell, products, intermediates, and individual substrates. The multipurpose automatic analyzers are modular in nature and the modules are united to give sets which are best for certain assays.

The important parts of the automatic analyzers are

- Sample collector which is used or the offline analysis
- Peristaltic pumps which are installed with tubings of diameter ranging from 0.1 to 3.9 ml/min.
- The sample is allowed to flow through the sampling tube and air bubbles are allowed to flow into the fluid column in order to create minute regular segments. The bubbles are used to prevent the dilution. Microorganisms and other particles are filtered prior to sample entering to the system. Continuous filter is used for filtering. It consists of a strip which will continuously be unwinding from the stock roll. Each junction of two solutions in the tube is followed by a stirring spiral which is made up of glass. If the sample contains impurities, then it is dialyzed first.

Ion selective electrodes

Ion selective electrodes (ISE) consist of electrochemical membrane which exhibits different permeability. The membrane is a layer which contains solid electrolyte or it can be an electrolyte solution in a solvent which is immiscible with water. The membrane will be in continuous contact with the electrolyte solution on both sides or it may be on one side. ISE generally contain internal reference electrode and in some cases, it is replaced by metal contact or by a semiconductor layer and an insulator. Ion selective electrodes helps to detect with high sensitivity, exact, rapid and is one of the best nondestructive method to measure ion concentrations in various media. They offer numerous advantages which are of practical use. They do not disturb the solution to be tested. They are mobile and can be cost effective. ISE can be used to measure minute volumes of culture. They are some limitations of ISE in terms of sterilization, its application in suspended particles and surface growth while using filamentous fungi.

Enzyme electrodes

The biological sensors utilize enzyme catalyzed reactions and are specific. Enzyme electrodes are the sensors which are used generally. It contains electrochemical sensor and a semi permeable membrane along with a intermediate layer. The intermediate layer contains enzyme. The function of this system is based on the substrate diffusion and the enzyme substrate reaction. The reaction might yield a product and it can be detected by an electrochemical sensor. The main limitation is that these electrodes are slower. This limitation made scientists to insert a semi permeable layer attached with the immobilized enzymes on the surface or to the interior of the membrane. After these alterations, the response is much faster than anticipated.

The sensors can be classified into three groups:

- The microbial sensors which contain immobilized cells and also an oxygen electrode. This is nothing but a amperometric respiration assay
- The microbial sensors which contain immobilized cells and also a fuel cell type electrode. This is a sensor which is an amperometric assay for measuring metabolites which are electroactive.
- The microbial sensors which contain immobilized cells along with membrane electrodes. These sensors are based on the potentiometric measurement of metabolites which are electroactive.

Mass spectrometry

This method is used to measure the substrate concentrations in any microbial process. It is advantageous because

- It allows the measurement of sample on a continuous scale and in aseptic conditions
- It permits measuring of various components at a time.

- This device is cheaper and can be used for measurement with good accuracy
- This device is attached with a computer system

Quadrupole mass spectrometer is a good choice for these types of measurements. Extra care should be taken while selecting the membrane which is very important and it affects the response time. The membrane should have good rigidity and permeability and the membrane needs to be sterilized readily. When analyzing quantitatively, there should be difference between mass transfer and the errors attached with the noticed spectrometric measurements. If the molar fractions of the components measured which are in the bioreactor and the inside the vacuum system should be same so that the problem night be minimized. If the membrane is utilized for the sampling, the molar fractions of the materials or the components on both sides of the membrane may differ. Calibration is another important problem which is to be addressed.

Physical environmental factors

There should be an extra care taken when selecting the construction materials since a trace of heavy metal whilst sterilization or even during culturing can adversely affect the growth of microorganisms and also decides the product formation. The optimal temperature for microbial culture ranges from 20 to 40 ^{0}C. If thermophilic microorganisms are used the optimal temperature ranges from 60 to 70 ^{0}C. If the selected microorganisms are psychrophilic in nature, then the optimal temperature should be maintained between 10 to 20 ^{0}C. If the temperature is more or less than 1 degree, then it might not affect the microbial growth. The measurements of temperature are done by resistance thermometers, thermistors, thermocouples and mercury thermometers. The glass mercury thermometer is installed inside a vessel with contact fluid. In this arrangement, the culture temperature and the thermometer reading can exhibit certain delays. The placement of thermometer should not be in direct contact with the nutrient medium so as to avoid damage to the thermometer. The direct reading thermometer is used widely and it is installed on the bioreactor and is depend on the bioreactor used. Due to their less voltage, the thermocouples are avoided. Thermistors should be calibrated regularly because resistance increases as the temperature rises almost in a non-linear way. The devices used generally are resistance thermometers which are stable. The temperature is controlled by certain heating elements and also with the help of coolant or a heating fluid through culturing devices. The pressure can be determined by with the help of Bourdon manometer. This manometer is installed below the air outlet filter. The filter may be maintained in dry conditions since wet conditions may affect the readings of the manometer. Good results are obtained by using diaphragm manometers if attached at the end of the Bourdon tube. The diaphragm is made using stainless steel and is arranged in between the two flanges. One of the flanges is installed above the lid of the bioreactor. The diaphragm can be easily sterilized. The pressure readings can be shown by pointer or by a scale.

Gas and fluid flow

The gas flow is determined by rotameters or flow meters or by diaphragms with calibrated throttle. The accuracy which is available by these flow meters is generally plus or minus 2 percent. The ratio observed with maximum and minimum flow rate is 10:1. The instruments can be sterilized with steam and they are mounted near the air filters. The accuracy of the measurements should be based in the pressure present in the piping. The measurements when higher flow rates are present can be done by using turbine flow meters. It is expensive but it gives good and accurate readings. The calibrated throttle diaphragms are generally used in industrial bioreactors. The main characteristic of these devices is its high rate of accuracy, free from pressure fluctuations, remote visualization, recording and regulation of flow rate, and low maintenance and investment costs. The diaphragm is installed at the air waste outlet so the risk of contamination is low. The flow of liquid is measured by various flow meters. It can be done by measuring the liquid weight and via dosage devices or pumps. The devices which are used are primarily depending on fluid volume measurements. They are simple and of low cost. The measurement of fluid mass is accurate because no dependency with temperature, viscosity, pressure, or any other relevant physical factors. The measurement of fluid volume is done by piston pumps, diaphragm and tube. The tube pumps can be linear or rotary and they are used in the pilot scale or production scale. By changing the tube diameter and the rate of tube compression the aseptic conditions are maintained. The quality of tube material is often important since the compression is done at varying degrees. The use of diaphragms is also of common practice while using lower flow rates. The fluid which is to be pumped can be separated from respective environment with the help of flexible diaphragm. The motion is migrated into the membrane through an auxiliary fluid with the help of piston or through direct method. The piston pumps which are to be used for sterile operation should be enclosed with a cover in order to protect the parts from direct contact and hence can reduce contamination. The main requirement is that all the measuring and dosage devices should be sterilizable by steam. The dosage devices are utilized in the process of medium addition, inoculum and also in regulating pH and defoaming in case of continuous cultivation.

Measurement of density, weight, and fluid volume

The medium or the culture volume present in the bioreactor can easily determine by measuring in non-aseptic conditions using calibrated ladder gauge and also via measuring the fluid height observed through an observation slit. Another important measuring device which can be used is a probe and it is based on the principle of capacitance. The weight of the content of the bioreactors of small scale can be measured by using scales and in case of large scale bioreactors tensometric balance equipped with semiconductor tensometers can be used. Constant weighing of bioreactors is essential in case of continuous cultivation in order to maintain culture volume at constant level. This is done in cases of higher flow rates and culture foaming. The culture broth density is measured by dividing the weight of the broth by its volume.

Regulation of foam

The formation of foam is an essential process if not common especially in any microbial process when utilizing complex natural raw materials. If foaming is uncontrolled, it can contaminate the culture fluid. The over flow of foam can reduce the culture volume in the bioreactor. The removal of foam is generally done by mechanical measures or mechanical foam breakers. The foam can also be removed by synthetic defoamers which reduces surface tension. Based on the composition of the medium, the microorganism the foams can be of light with large air bubbles or they can be heavy with tiny air bubbles. The antifoam agents are generally added to the medium during the first stage of cultivation whereas the defoaming agent is added during the process of cultivation.

The detection of foam can be done by two sensors and these sensors are based on capacitance and conductance. The sensors which are based on the conductance contains probe made up of stainless steel which function as electrode and the lid of the bioreactor can act as earthing. If foam reaches the uninsulated end present in the electrode a unit gets activated and the defoamer dosage gets setup. Some of the sensors utilize two probes. One probe is placed at the higher place and the other probe is placed at the lower portion of the bioreactor. The probe present in the upper end activates the defoamer dosage and the probe placed in the lower portion of the bioreactor is used to cut off the process. The probe which uses capacitance is basically a metal rod which is enclosed with Teflon layer. The probes used are expensive and are most reliable because of fewer disturbances. The defoamers should be used in minute amounts because they decrease the rate of oxygen transfer and can make the product isolation complicated. A lag is present between foam subsidence and defoamer addition.

Measurement of impeller

The speed of the impeller while mixing media is dependent on the size of the bioreactor. It will be in the range of 100 to 2000 per minute and the speed decreases with the increase in the volume of the bioreactor. The drive of the impeller is regulated by DC motors which are equipped with thyristor speed regulation. The speed of the impeller can be regulated by a potentiometer and can also be measured by contact less induction system. The induction sensor is installed over the engine shaft and is capable of giving information regarding engine revolutions. The measurement of the power input of the impeller and the torque are so important for the scale up process and also in determining the power to be consumed by the impeller. The large-scale bioreactors are installed with wattmeters and this arrangement is not advisable in case of small scale bioreactors due to energy losses. The small-scale bioreactors are fitted with extensometers or dynamometers. The torque meter or the torque dynamometer measures the torque between the impeller and the engine shaft. The shaft is located in a bearing which is free revolving with the help of coil or spring. The torque gives deviation of the coil or the spring and this can be read on the scale. The building of the extensometer is dependent on the variations of conductivity present in the resistance wire and it is due to the transversal or the mechanical load. The extensometers are installed on the shaft and inside the bioreactor. Four tensometers are fitted at 45^0 angles to the axle. With this angle, the shear deformation is higher.

Viscosity

The measurement of viscosity is crucial since it determines the morphology and cell growth and it also gives the physical characteristics of the microbial culture. The cultures which contain unicellular organisms behave like Newtonian fluids and the multicellular microbe behaves as non-Newtonian fluid and the viscosities changes as per different shear velocities. The viscometer should allow the determinations at required shear velocities. The simple viscometers cannot differentiate between pseudoplastic and plastic fluids. Couett viscometer is installed with rotating coaxial cylinder and well-defined shear velocities. The important factor which is taken into account is the shear which is fluctuating in the space present between the cylinders.

Chemical environmental factors

The chemical environmental factors are very important to be measured. The important factors are pH, rH, partial pressure and concentration of the dissolved oxygen, partial pressure and concentration of dissolved carbon dioxide, and the concentrations of other gases such as carbon monoxide, hydrogen and methane. The concentrations of carbon and oxygen are also measured in the air outlet. The measurement of concentrations of the products, intermediates and substrates is also crucial. The measurements are generally conducted by ion selective electrodes and also the electrodes which used immobilized biocatalysts. These are used for on line measurement on continuous basis and also usage of mass spectrometric methods is not uncommon.

pH and rH

The pH values are measured and later recorded and sometimes measured followed by regulation in the range of pH from 2 to 12. These systems contain sterilizable electrodes, measuring instruments, regulation and recording, reservoirs for alkaline and acid solutions and also pumps. The solutions are injected slowly often far from the electrodes in order to make the process as precise as possible. The titrant doses should be adjusted to match the feed pulses and their length and the pH regulation. The redox potential is determined using devices which are similar to the pH meters. The platinum electrode is used instead of glass. The redox potential can be controlled by injecting sodium thioglycolate, ascorbic acid, cysteine, oxygen or nitrogen. The redox potential is not measured regularly.

Dissolved oxygen

The dissolved oxygen is one of the important factors since oxygen levels are required for the microbial growth. A sudden variation in the dissolved oxygen level can reflect the sharp change in the cell metabolism. The partial pressure and the concentration of the dissolved oxygen present in the fluid can be detected by using oxygen electrode. The anode is made up of platinum wire and the cathode is the ring made up of silver oxide and silver. The electrolyte is a gel made up of cellulose and contains potassium chloride. This membrane can be fabricated using stainless steel mesh and Teflon is also used. The response time of the electrode is less than 10 seconds and the accuracy is more or less than 1 degree Celsius. The temperature which can be measured ranges from 5 to 50 ^{0}C. The electrodes are made up of stainless steel or glass and can be sterilized with steam. The signal from the electrode passes from an amplifier and then monitor and finally recorded or read my line recorder. These electrodes can detect the partial pressure of oxygen and also actual oxygen. The oxygen concentration is directly proportional to the partial pressure. The solubility of oxygen gets decreased if the dissolved substances increase. The accuracy with the oxygen electrodes is affected by several problems due to inhomogeneity. The oxygen dissolved in the broth can get diffused through the wall of the tube into the nitrogen stream and the concentration is detected by oxygen analyzer. The dissolved oxygen can be measured by using mass spectrometry. The dissolved oxygen concentration can be controlled by altering the volume coefficient of mass transfer or by altering oxygen concentration. The volume coefficient of the transfer of oxygen can be altered by modifying impeller speed. If the dissolved oxygen level is lowered below critical level, the speed of the impeller should be increased. If the tube outlet is lower inside the culture fluid, there will be longer residence time of air bubbles and higher oxygen transfer. The regulation of the dissolved oxygen concentration at the inlet can be done with two methods. One is by changing the oxygen to nitrogen ratio in the gas which is to be pumped maintained at constant gas volume. The other method is by altering the quantity of the air which is to be fed into the constant flow of nitrogen. The former method demands changes in the speed of the impeller and the latter method demands constant impeller speed. The oxygen level is controlled by the rate of carbon source added into the bioreactor.

Measurement and regulation of gases

The dissolved carbon dioxide concentration is measured by potentiometric sensor. In this sensor, the internal probe is just a glass electrode immersed in sodium carbonate solution. The equilibrium pressure of the carbon dioxide and its concentration outside can be attained in the pores of the membrane and ideal carbon dioxide concentration can be established in the solution which is encircling the glass electrode. The microporous membrane has allowed many to construct gas probes which are commercially used. The response time of electrodes and gas sensors is based on the gas diffusion via the microporous membrane. The gases which are dissolved in the culture broth are hydrogen sulphide, nitrogen, methane, hydrogen and carbon dioxide. The measurement of these gases can be done with quadrupole mass spectrometer with high precision and accuracy.

Determination of gases

Many gases stated above are released in the form of products or sometimes consumed in the form of substrate. In either case, the concentrations of the respective gases in outlet and outlet of the bioreactor should be measured. Oxygen can be determined by using paramagnetic oxygen analyzer. The oxygen concentration can be measured by bubbling the gas via a water column in which the dissolved oxygen is determined at constant temperature by using oxygen electrode. The carbon dioxide content is determined by the infrared gas analyzer. Methane gas can be detected and measured by using a biosensor which has immobilized cells belonging to Methylomonas flagellate. It contains immobilized cells along with reference reactor and also two oxygen electrodes.

Measurement of substrate and product

A chemical can act as a product or the substrate at the same time. This condition occurs in case of amino acids, organic acids and alcohols. Commonly used saccharides are lactose, sucrose and glucose. The good and sensitive technique is the use of enzyme electrodes. The enzyme electrode which was used is glucose oxidase electrode in the glucose assay. This is improvised by using immobilized glucose oxidase. The immobilized cells of *Pseudomonas fluorescens* in a collagen membrane can be used as sensor and these cells have the capacity to utilize glucose. It uses collagen membrane and it contains bacteria. Teflon membrane which is permeable to oxygen, lead anode, platinum cathode and alkaline electrolyte. The concentration of the sucrose can be measured by enzyme thermistor or the yeast cells which are immobilized on the porous glass.

The alcohols which are to be determined are butanol, methanol and ethanol. Ethanol can be measured by a biosensor, quadrupole mass spectrometer, semiconductor sensor, enzyme electrode and ISE electrode. Methanol can be measured by bioelectrochemical cell. It contains an anode and a cathode chamber. These two are separated with a membrane. The anode chamber consists of methanol dehydrogenase which is extracted from pseudomonas and the buffer is maintained at 9.5 pH. The nitrogen gas is allowed to pass over the solution and it is magnetically stirred. The current which is being generated in the cell is determined by the potential difference. This bioelectrochemical cell can be developed by adding an immobilized enzyme or the cells. The concentration of methanol can be measured by mass spectrometer. Butanol can be measured by using acetone-butanol fermentation. Some of the methods include mass spectrometry and headspace gas chromatography. Acetic acid can be measured by quadrupole mass spectrometer and also by biosensor. Formic acid can be measured by using fuel cell tape electrode. This apparatus contains an electrode, Silver dioxide, platinum anode and Teflon membrane. The membrane contains clostridium butyricum and it is placed on the Teflon membrane and also covered with porous Teflon membrane. The unit which is used for measuring contains a microbial sensor, chart recorder, mV amperometer and indicator.

Hydrogen is produced from the formic acid. It gets penetrated through the Teflon membrane and later oxidized on the platinum anode. The relation between the formic acid and signal current is linear. The least detectable concentration is 19mg per liter. The sensor has the stability of 20 days. Lactic acid can be measured by an enzyme electrode. It contains enzymes cytochrome b2 and yeast lactate dehydrogenase. Butyric acid can be measured using quadrupole mass spectrometer. The amino acid concentrations can be determined using high performance liquid chromatography and automatic analyzers. It is advisable to use biosensors. Some of the characteristics of the enzyme electrodes used in amino acid assays are

- D-amino acids can be a substrate with D-amino acid oxidase as an immobilized enzyme with a stability of one month
- D-amino acids can be a substrate with L-amino acid oxidases as an immobilized enzyme with a stability lesser than one month
- L-Asparagine can be a substrate with asparaginase as an immobilized enzyme with a stability of one month.

The Nitrogen sources can be measured by ion selective electrodes. Generally used electrodes are Orion ISE equipped with digital pH meter. Ammonium electrodes generally use potentiometric detection. This is made by the combination of gas permeable membrane and glass electrode. The measurement generally occurs in presence of alkaline pH. Amines are the main contaminants in this process. The nitrification bacteria namely *Nitrosomonas* uses ammonium ions as the source of energy and the assimilation of this ions is linked with the consumption of oxygen. This bacterium is utilized in its immobilized form which can be used for amperometric determination of these ions. This system comprises of an immobilized bacterium equipped with an oxygen electrode. Ammonia which is in gaseous state diffuses via a permeable membrane and it gets assimilated by the bacteria. The biosensor for the nitrate measurement is dependent on *Nitrobacter* species equipped with an oxygen electrode. The immobilized cells are arranged in rings and they are placed over the oxygen electrode and are covered with another membrane made up of Teflon which is permeable to gases. Antibiotics are measured by automatic analyzers. In this test, the microorganisms which are sensitive to a particular antibiotic are cultured continuously in a chemostat and the microbial culture is supplied with particular to be tested. The concentration is measured by the decrement in carbon dioxide production. Cephalosporins and penicillin can be measured by biosensors. The biosensors for measuring cephalosporin are dependent on Citrobacter freundii which contain cephalosporinase. This enzyme liberates hydrogen from the cephalosporins. These cells are immobilized in a membrane made up of collagen. The biosensor for measuring penicillin contains an enzyme electrode. The electrode contains immobilized penicillinase. This enzyme has the capacity to hydrolyze penicillins to penicillanic acid. If the pH drops, it is due to the acid formation.

Biological environmental factors

The key quantity for regulating the microbial processes is the cell mass concentration. These measurements can be done by using direct and indirect methods. The microbial concentrations can be measured by biomass count or simply by cell count. The simplest method is by making a cell count using cell counting chamber or by plating the microbe on the agar Petri plates. The modern methods include particle analyzer, laser flow microfluorometers, flow cytometer and coulter counting. The microorganisms are grown and plated on above the other and were passed under the beam of light in laser fluorometer. Helium-Neon laser is used to obtain infra-red radiation and continuous Argon laser is used to achieve UV light. In flow cytometers, the microorganisms can be differentiated from other inert particles due to their variations in dielectric properties. The biomass concentration can be determined by indirect and direct methods. In direct method, the microorganisms are separated from the culture broth either by the process of filtration or by centrifugation. The indirect procedures are simple to use and are faster. Densitometry is one of the methods which are widely used. If the culture fluid is free from solid particles, the optical density can be determined by using spectrophotometry. These values are proportional to the particle number for unit volume. This method holds well if the particle size is constant and the optical density is managed at 0.5 or less than that. Biomass can be measured by electrochemical sensors which is based on the principle same as fuel biocell. It combines the consumption and production of electroactive substances with the help of microorganisms.

This is due to the electrochemical interaction between microorganisms and electrode. Chemical determination of DNA is another method which can be used to measure biomass concentration in suspension or the solid. Bioluminescence ATP assay with the cells is another good method. These assays are based on the DNA and ATP content of the cells.

Determination of microbial population and homogeneity

The ratio of dead cells to the live cells in a microbial population can be measured microscopically using some staining techniques. Rapid methods such as biosensors and flow cytometry are being used for this purpose. The distribution analysis is performed by laser flow microfluorometry ad photon correlation spectroscopy.

Determination of intracellular components

Automatic analyzers are used to measure the biomass composition. DNA, RNA and proteins of the cells can be measured by injecting fluorescence with help of dyes.

The computer installed in the bioreactor can help in the following tasks

- Acquisition, sorting and recording of data
- Reduction and analysis of data
- Control of process
- Optimization of process

The process control is done in two methods

- The computer can adjust the constants
- The regulators can be replaced by computers

The process optimization can be in three approaches

- Feedback search system
- Based on accurate model of the process
- Usage of adaptive method

REFERENCES

1. Ackland, M. R., Manvell, P. M., Bean, P. G. (1984) Biotechnol. Lett. 6, 137.
2. Armiger, W. B. (1985) Instrumentation for monitoring and controlling bioreactors. In: Comprehensive
3. Biotechnology 2, New York.
4. Barker, A. S., Somers, P. J. (1978) Enzyme electrodes and enzyme based sensors. In: Topics in
5. Enzyme and Fermentation Biotechnology 2, Chichester.
6. Barker, A. S., Emery, A. N., Novias, J. M. (1971) Process Biochem. 6, 11.
7. Benson, C. M., McDougal, D. C., Coffey, D. S. (1984) Cytometry 5, 512.
8. Bull, D. N. (1983) Ann. Rep. Ferm. Proc. 6, 359.
9. Bull, D. N. (1985) Instrumentation for fermentation process control. In: Comprehensive Biotechnology
10. 2, New York.
11. Clarke, D. J., Blake-Coleman, B. C., Carr, R. J. G., Clader, M. R., Atkinson, T. (1986) Trends Biotechnol.
12. 4, 173.
13. Comberbach, D. M., Scharer, J. M., Moo-Young, M. (1985) Enzyme Microb. Technol. 7, 543.
14. Cordonier, M., Lawny, F., Chapot, D. (1975) FEBS Lett. 59, 263.
15. Corrieu, G., Blach6re, H., Geranton, A. (1974) Biotechnol. Bioeng. Syrup. 4, 607.
16. Cox, R. P., Jensen, B. B., Joergensen, L., Degn, H. (1984) Microbiol. Sci. 1,200.
17. Cullen, L. F., Rusling, J. F., Schleifer, A., Papariello, G. J. (1974) Anal. Chem. 46, 1955.
18. Dairaku, K., Yaman6, T. (1979) Biotechnol. Bioeng. 21, 1671.
19. Dobry, D. D., Jost, J. L. (1977) Ann. Rep. Ferm. Proc. 1, 95.
20. Doerner, P., Lehmann, J., Piehl, H., Megnet, R. (1982) Biotechnol. Lett. 4, 557.

21. Dolbeare, F. A., Vanderlaan, M., Phares, W. (1980) J. Histochem. Cytochem. 28, 419.
22. Grayson, P. (1969) Process Biochem. 4(3), 43.
23. Guilbault, G. G. (1976) Handbook of Enzymatic Methods of Analysis, New York.
24. Harmes, C. S. (1972) Dev. Ind. Microbiol. 13, 146.
25. Harrison, D. E. F., Harmes, C. S. (1972) Process Biochem. 7(4), 1.
26. Hatch, R. (1982) Ann. Rep. Ferm. Proc. 5, 291.
27. Hatch, R., Wilder, C., Cadman, T. W. (1979) Biotechnol. Bioeng. Symp. 9, 25.
28. Herbert, D., Phipps, P. J., Tempest, D. W. (1965) Labor. Pract. 14, 1150.
29. Hikuma, M., Kubo, T., Yasuda, T., Karube, I., Suzuki, S, (1979) Biotechnol. Bioeng. 21, 1845.
30. Hikuma, M., Kubo, T., Yasuda, T., Karube, I., Suzuki, S. (1980a) Anal Chem. 52, 1020.
31. Hikuma, M., Obana, H., Yasuda, T., Karube, I., Sizuki, S. (1980b) Anal. Chim. Acta 116, 61.
32. Hjortso, M. A., Dennis, K. E., Bailey, J. E. (1985) Biotechnol. Lett. 7, 21.
33. Hospodka, J. (1966) Biotechnol. Bioeng. 8, 117.
34. Humphrey, A. E. (1977) Process Biochem. 12, 19.
35. Jefferis, R. P. (1975) Process Biochem. 10(4), 15.
36. Karube, I., Suzuki, S. (1983) Ann. Rep. Ferm. Proc. 6, 203.

CHAPTER 8

BIOLOGICAL CATALYSTS

INTRODUCTION

The process of immobilization numerous techniques which include animal and plant cells, microbial cells, cell organelles and bonding enzymes. It can be described as immobilization or binding of bioactive molecules such as antibiotics, antibodies, antigens and enzymes with a phase which contains dissolved substrates or products.

Immobilized enzymes

The enzymes are considered as natural catalysts. It is said that approximately 2000 enzymes are characterized and they form the central core of life in microorganisms, plants and animals. The enzymes can function under very mild conditions in presence of neutral pH under normal pressures and temperatures. High specificity can be achieved by the enzymes. In enzyme catalysis, there will be no damage to either reactants or the products. The primary limitation of using these enzymes is their limited availability. Enzyme immobilization is used if

- The enzymes are exported from the cells that is extracellular
- The high molecular weight substrate is used and there will be a limitation since these molecules cannot enter the cell
- The enzyme may not be of microbial origin and should be extracted from animal or plant tissue prior to immobilization

- The product must not contain even traces of other substances which are released by immobilized cells
- The enzyme immobilization is advantageous due to economic reasons and therapeutic reasons.

The immobilized cells offer the below advantages

- The enzyme stability can be increased
- The costly affairs such as extraction followed by purification can be avoided
- The conversion of complex enzymes can be achieved
- Biosynthesis of different substances is possible
- The economics of the immobilized process is also improved

If the product is degraded enzymatically or synthesized and if the process is independent of the cell growth then the process of immobilization does not require cells which can grow. The immobilization process can use the cells which can be killed without damaging the enzyme apparatus. The cells need to contain excess amount of enzyme and so the overproducing strains are to be used. The substrate is supplied in a way that the enzymes modify them in presence of optimum temperature and pH. The main problem with the immobilized cells is that there will be less diffusion velocity of products and substrates. Therefore, it is essential to enhance the size of the reactor. The immobilized cells can also possess other enzymes which can act on the substrate and may produce toxins. There are some processes which can stabilize the enzyme desired and the other enzymes are inactivated by heating to 80 ^{0}C for some minutes. The permeability of the cells can be enhanced by autolysis, lauryl sulphate, cetyltrimethylammonium bromide and toluene prior to immobilization.

If the cells of the death phase are used, the cells are immobilized in a way that the cells immobilized can no longer have growth activity. These cells can sustain certain metabolic activity. Since the cells have no capacity to proliferate, they can yield the desired products using the added substrates. If the microorganisms are still growing, then they would consume some substrates which is not useful. The immobilized cells can get multiplied in a synchronous way. The daughters' cells retain almost the same generation time same as the immobilized cells. The immobilized growing cells provides some advantages such as

- The cells which can be immobilized can be low
- The cell concentration in the carrier is greater than that in the liquid medium
- There will be greater productivity since a dense layer is produced on the top of the carrier whilst growth
- The stability of the production is greater since the cells get multiplied while operation
- The rate of dilution can be independent of the rate of growth

Co-immobilized biocatalysts

These biocatalysts are of good choice when the course of reaction is of biotransformation and or simple biosynthesis. The process of co-immobilization is done by using cells or enzymes, coenzymes or the enzymes with substrates which are insoluble, high density substances which help in biocatalyst sedimentation and also substances which serve for pH adjustment. Some of the possibilities in preparation of this type of biocatalysts are

- Two or more enzymes
- Enzymes which also contain coenzymes
- Two or more types of living cells
- Two or more types of life less cells
- A mixture which contains both non-living and living cells
- Non-living cells associated with enzymes
- Cells or enzymes with a solid substrate
- Cells or enzymes with high density particles
- Cells or enzymes with materials which affect the pH

These biocatalysts can be termed as second generation immobilized biocatalysts.

Basic prerequisites

The immobilized biocatalyst can be prepared by considering certain prerequisites

- Biocatalyst which is suitable

- Suitable gel or carrier
- Suitable technique for immobilization

Whatever the procedure of immobilization is employed, there should be a confirmation that the immobilized biocatalysts have their enzymatic activity. The immobilization must yield biocatalyst particles which can be separated easily from the solution and they should be resistant to abrasion and are mechanically firm. The biocatalysts must have optimum size and surface area. Maintaining of aseptic conditions while immobilized biocatalyst preparation and later in the bioreactor process is important. The agents which are used for sterilization are radiation, antibiotics, sorbic acid, ethanol and formaldehyde. The storage of immobilized catalysts can occur in wet or in dry state. The below properties must be assessed during the selection of support materials and carrier for immobilization.

- The chemical functionality or the chemical stability of the active group
- pH, temperatures and relevant environmental factors
- Resistance to microbial attack
- The internal surface area and pore volume
- The amount of water retained and the hydrophilicity of the carrier
- Based on the binding capacity of the catalyst and also its mechanical characteristics

The materials which are used for immobilization is diverse. The materials which are used for this process are classified into two groups. They are

- The carriers or materials to which the biocatalyst can be attached chemically, physico-chemically or physically
- The materials in which these biocatalysts can be entrained or entrapped

The substances to which these biocatalysts can be bound are plant residues, dextran and cellulose. In addition to this, several polysaccharides, and proteins such as albumin and gelatin. Some of the inorganic carriers are titanium oxides, zirconium, aluminium, porous steel, sand, ceramics and glass. Synthetic polymers which can serve for this purpose are polyurethane, resins which are light cross linked, polyvinyl chloride and polyacrylamide. Natural polymers include collagen, agar-agar, and carrageenan and calcium alginate. The encapsulation of the biocatalysts can be performed if the material chosen can allow the production of semi permeable polymer membrane.

Techniques for immobilization of enzymes

Nelson and Griffin were the pioneers in adsorbing invertase enzyme onto the aluminium oxide and charcoal. Grubhofer and schleith succeeded in adsorbing carboxypeptidase, ribonuclease, pepsin, amylase onto diazotized ployamidostyrene. Bar Eli and Katchalski has immobilized trypsin to the copolymer of leucine and p-aminophenylalanine. Manecke and Gunzel have immobilized non-hydrolytic enzymes like alcohol dehydrogenase onto the polymers which are synthetic. Some of the techniques used in the process of enzyme immobilization are

- Adsorption of enzymes onto the surface of the carrier
- The formation of the covalent bond between surface of the carrier and the enzyme
- Cross-linking or copolymerization of enzyme molecules
- The entrapment of enzymes in the matrix
- The anchoring or the encapsulation of the enzymes to the membrane structure

Adsorption

The physical adsorption is an interaction which is between enzyme molecules and carrier. Due to the weak interactions, the bond is highly sensitive and the enzymes can be easily desorbed by slight variations in pH, reactant concentrations, temperature and ionic strength. Therefore, the biocatalyst which is stable is hard to achieve. The instability is inherent in this process and hence it is not advisable in the immobilization process. One of the best alternative methods lies in using ionic bonds. The ionic bond can provide strength greater than the physical adsorption. The biocatalyst which is formed by this process exhibits more stability at optimal pH and sufficient ionic strengths maintained. On changing ionic strength and pH, the enzyme can be liberated from the carrier and may possibly be reused. The metal bonds and chelation can be used to attach enzymes with the surface of the carrier by agents such as titanium oxide.

Mild conditions of binding are employed as the activation followed by neutralization of the carrier is done before the enzyme gets into contact. The weak interactions between carrier and the biomaterial is a part of physical adsorption. Ligand exchange or replacement is the underlying phenomena in the process of adsorption. Carboxylation or amination of the carrier is employed to obtain a good and uniform product and it diminishes the inhibitory reactions of the titanium molecule. The process of immobilization using adsorption is explained by Goldstein, Manecke, Chibata and Messing.

Covalent bonding

The immobilization of the enzymes with the use of covalent bonds is an easy method as this method can be used in various pH and ionic strengths. The limitation of this step is that this method requires several steps. The chemical bond can be yielded using various functional groups which is of a protein moiety of the enzyme structure. Some of the functional groups are hydroxyl groups, imidazole group, sulphydryl group, and alpha-beta and gamma carboxy groups. The carriers which are suitable are proteins, carbohydates and other inorganic materials.

The binding of a particular enzyme with the carbohydrate carrier is done by binding cyano group of cyanogen bromide with two other hydroxyl groups of the surface and binding the cellulose carboxyl group to the carbodiimide. Some of the polyfunctional agents are used to bind enzymes or proteins to a carrier. Diazobenzidine binds to the tyrosyl groups. Glutaraldehyde can form Schiff base with the help of amino acids. The inorganic carriers used in this process contain silan bonds. This method involves the silan linkage. It can be between gamma aminopropyl triethoxysilan on inorganic surface and onto the enzyme amino group. The amine in the terminal end can be attached to the enzyme through a polymerized isocyanate or by a bifunctional agent.

Cross linking

This method is dependent on the intermolecular bonds between polyfunctional agents and enzyme molecules. The enzyme molecules can bind to other molecules too. The activity of enzymes of immobilization will be based on cross linking. If the cross-linking agents do possess higher concentrations, then there would be formation of several bonds. If the cross linking is more, then there would be least penetration of the substrate and hence hindering the overall activity.

Entrapment

Enzyme trapping is a physical immobilization method in the matrix network. The enzyme cannot be a chemical component of the matrix. The matrix can be made up of polyacrylaide, alginate, agar, collagen and also starch. The entrapped enzyme can easily be escaped into the outer environment. This is due to the presence of the large pores in the gel. If large substrate molecules are present the diffusion would be via pores. Free radicals which form due to the gel polymerization can inactivate enzyme. The escape of enzyme may be prevented with the help of cross linking the entrapped enzyme via a bifunctional agent. These techniques are primarily used for the process of immobilization.

Encapsulation

The encapsulation or the process of encapsulation of any enzyme is based on the semipermeable polymeric membrane and this membrane protects the enzyme from being inactivated. This process includes polymerization of the interface, solvent evaporation, and separation of the phase. The encapsulated enzymes are used for low molecular substrates. The percentage of the immobilization is highest in this method. The materials which are used as constructing membranes are collagen, ethylcellulose, acetates and cellulose nitrates. The semipermeable membranes are like flat plates, hollow fibers, or dialysis tubes. This method is simple and the enzyme will remain active in the solution in its natural form. If the diffusion velocity of the substrate is increased and altering the product molecules can result in the loss of enzyme activity. This is observed in the membranes with higher rate of porosity.

Cell immobilization

The techniques for the process of cell immobilization can be classified into three groups. They are

- Cell immobilization without using a carrier
- Cell immobilization with a carrier

- Cell immobilization with the help of encapsulation or by trapping inside the polymers

Cell immobilization without a carrier

This immobilization technique does not require synthetic or natural carriers. It requires stabilization or immobilization of the enzymes present in the cell or the aggregated cell particles. The best way to stabilize the enzymes present in the cell is to heat the cells to 80^{0}C for hours. The important factor to the considered is that the enzyme behavior in the higher temperatures. The undesired enzymes such as proteinases can be denatured due to this heat treatment. If the cell is exposed to the citrate solution, then there will be reaction with chitosan which is present in the cell walls and followed by heating until it is dried. The pellets of the filamentous microorganisms such as actinomycetes or fungi can be described as form of the immobilized cells. The immobilization of the cells can be done via the process of flocculation with the help of the agents like anionic or cationic polyelectrolytes. Poly amine and carboxy substituted polyacrylamide are the examples of these agents. Some of the other agents are manganese, ferrous and ferric agents, calcium, phosphates of magnesium, sulphates of magnesium, hydroxides or the oxides of magnesium and calcium and mineral hydrocolloids. The cells which are flocculated are dried and then processed to produce particles of the size desired by the user. The cells which are immobilized using this method have low mechanical resistance.

Another key method is the cross linking of intracellular materials and cell agglomerates such as pellets or the flocs. This will stabilize the cell against the action of autolysis. The cross-linked particles exhibit mechanical firmness and sedimentation and so the enzyme activity is enhanced. The elasticity and the mechanical strength of the cell aggregates can be increased further using flocculation agent. The flocculation agent can be a cationic agent and it reacts with the cross-linking agents. The cross linking can be done using polyfunctional agents and they can penetrate easily into the cell and can react with cell wall, cytoplasmic membrane and cytoplasm. The polyfunctional agents are gultaraldehyde, pyrimidine, pyridine, acriflavine, acridine which are diazo derivatives, chlorinated triazines, 2,4 toluenediisocyanate and dialdehydes of dicarbonic acids. The treatment of cell aggregates and cells with certain agents can result in reducing enzyme activity. The main reason for the loss of activity is due to enzyme deactivation with the help of polyfunctional cross linking agents. In order to restore the ability of diffusion, the linked cells are allowed to be permeated by lauryl sulphate, cetyltrimethylammonium bromide and toluene.

The permeability of the biocatalyst can be enhanced if the cells to be used were autolyzed and disintegrated partially. The treatment liberates the contents of the cell and the cell post cross linked with 2,4 toluenediisocyanate. The biocatalysts are having high stability and they do have good mechanical and sedimentation properties. It is possible with the double cross linking of the cell components and the individual cells in first and the final steps of the reaction. A good preparation can be yielded by immobilization procedures which are water soluble and chemically reactive polymers and do have higher molecular weights.

Immobilization of cells with a carrier

This method of cell immobilization is done using coordination covalent bonds, chemical bonds, ionic bonds, physicochemical linkages or the polar linkages and physical adsorption. The carriers which can be used for this process are inorganic materials such as bentonite, ceramics, chips, granules, clays, sand and glass. The organic materials used in this procedure are ion exchange resins and cellulose. Hattori and Furusaka were the first to use such a type of immobilization. In case of physical adsorption, the cells are mixed with the mixture of carrier and this can be of any material which can react with the cell. Therefore, the binding is weaker and the cells get escaped from the surface of the carrier and it can undergo the process of autolysis. The activity of the enzymes is too low. The adsorption of any microbe depends on the carrier to be used.

The cell immobilization with the help of the chemical linkage can be done by the mixing of cells with the carrier which is already activated by the chemical means or modified in a way that the functional groups can be available to be reacted with the cell. The carriers which are suitable with this process are chemically reactive and they are organic polymers and in some cases monomers. Some of them are epoxide, aldehyde groups, with methacrylate and glycylmethacrylate. Some of the other groups of carriers are polymeric substances with hydroxides of ferric, zirconium, titanium. These biocatalysts exhibit good sedimentation and other hydrodynamic characteristics and can have higher rate of enzyme activity. The important limitation of this process is that it demands high costs. The microorganisms which are produced as wastes can also be used as carriers because they are cheap. The cell walls and cells do contain teichoic acid, mannan, chitin, chitosan, alpha D glucopyranosyl residues and other polysaccharides along with certain amino acids which have the capability of binding with the cells and enzymes. The stability of the cell aggregates and the cells which are immobilized wither chemically or physically can be enhanced by deploying cross linking properties in the cells and also by using the polyfunctional enzymes. The rate of the diffusion can be increased by the permeabilization and followed by cross linking.

Trapping

The process of the entrapping into the process is the good method especially when living cells are used. The substrate diffusion into the cells is easy and the cells are capable to grow. This technique not only immobilizes microbial cells but also plant and animal cells. The immobilization materials are basically synthetic polymers such as polyurethane, photo cross linked resins, polyvinyl chloride and polyacrylamide. The natural materials used are collagen, cellulose, gelatin, agar, carrageenan and calcium alginate. By this process the cells retain activity for higher periods. Cheetham, Imber and Isherwood have observed that the immobilized bacteria can be 350 times more stable than the free bacterial cells.

The biosynthesis of the catabolite products can be done with the living immobilized cell. The catabolite products are being produced by using dead or non-growing cells which are immobilized. The main use of the immobilized catalysts is that the process can be done for extended periods of time than the free cells. It can be for hundreds of days in continuous and semicontinuous operations. The genetic stability is enhanced in the state of immobilization. The cells which are immobilized are devoid of the mutations but it is not the case with the free cells. The ageing process however, affects the cells both physiologically and morphologically. In many productions, there is no substantial effect on the synthesis of the products. In some cases, the rate of permeability is affected in the plasma membrane and the cell wall and cell organelle reorganization do occur. Some of the researchers showed that the use of spores in the process of immobilization is beneficial since they are heat resistant and they can be sterilized easily. The species and genera of the micrrorganisms where the spore is used for immobilization is considered as important. Kren found that the immobilized conidiophores of *Claviceps fusiformis* with the help of alginate gel produced lower yield to that of its counter parts. Kopp and Rehm have found that there is no difference in the immobilized spores to that of vegetative cells in case of *Claviceps purpurea*. In case of free cells, the type of the inoculum and its age is crucial for immobilization to take place. Inoculum culture media and its age is also an important parameter to be considered prior to immobilization process. The composition of the medium for production and the culture medium for free cells differs specially while producing secondary metabolites. Lower concentration of substrates is used in this process since higher concentrations are usually unnecessary. The medium composition is crucial in secondary metabolite production since there will be inhibition of the biosynthesis. The rate of the biosynthesis will be directly proportional to the concentration of the cell in case of immobilized biocatalyst. In some cases, the accurate optimal ratio varies with the process it undertakes. Another important factor is the supply of the oxygen to the immobilized cells. The effect of oxygen can be described as the pellet growth. The deficit of the oxygen will result in inhibition of the respiration of the cell. Then,the experiments will be done using oxygen donors and it increases the oxygen availability for the cells. This technique is feasible technically but it is not profitable especially while going for long term operations. The rate of proliferation and the metabolic reactions in case of immobilized cells can affect the metabolite production both negatively and positively. The undesired metabolites are either overproduced in expense of the metabolite desired or the desired metabolite is produced with least production of undesired product.

Co-immobiized biocatalysts

The co-immobilized biocatalysts are the systems that will be there in a single particle cells or enzymes, mixture of cells and enzymes, cofactors with cells or the enzymes and biocatalysts which are there on solid substrates. The process of co-immobilization can be done using all the techniques described above. The choice of method is based on the purpose to which the method of co-immobilization is said to be used. Natural co-immobilization occurs in lakes, streams and water. The population of selected microorganisms and protozoa can be used as immobilized living cells for the process of anaerobic and aerobic water treatment and also in removal of non-toxic and toxic wastes. Co-immobilized biocatalysts are used with the substrate mixtures where no individual microorganisms will be capable of racting on the substrate. The important application is seen in the multi stage enzyme synthesis when the complex substances are used.

REFERENCES

1. Charles, M., Phillips, J. (1985) Combined immobilized enzyme/cell systems. In: Comprehensive
2. Biotechnology 2, New York.
3. Cheetham, P. S. J., Imber, C. A., lsherwood, J. (1982) Nature 299, 628.
4. Chibata, I. (1978) Immobilized Enzymes, Research and Development, Tokyo.
5. Chibata, I., Tosa, T. (1977) Adv. Appl. Microbiol. 22, 1.
6. Chibata, I., Tosa, T., Fujimura, M, (1983) Ann. Rep. Ferm. Proc. 6, 1.
7. Chibata, I., Tosa, T., Takamatsu, S. (1984) Mierobiol. Sci. 1, 58.
8. (~ulik, K., Sikyta, B., Slezfik, J., Palkoska, J., glechta, J., B~e(~ka, A. Herold, M., Ho~f~lek, Z. (1964) Czechoslovak Patent 113 908.
9. D6ppner, T., Hartmeier, W. (1984) Starch 36, 283.
10. Durand, G., Navarro, J. M. (1978) Process Biochem. 13(9), 14.
11. Fukui, S. (1983) Immobilized biocatalysts. In: Microbial Utilization of Renewable Resources 3,Osaka.
12. Goldstein, L., Manecke, G. (1976) Appl. Biochem. Bioeng. 1, 23.
13. Grubhofer, N., Schleith, L. (1954) Z. Physiol. Chem. -- Hoppe Seiler's 297, 108.
14. Hackel, U., Klein, J., Megnet, R., Wagner, F. (1975) Eur. J. AppL

Microbiol. 1, 291.

15. Hartmeier, W. (1985) Trends in Biotechnol. 3, 149.
16. Hartmeier, W., Heinrichs, A. (1986) Biotechnol. Lett. 8, 567.
17. Hattori, T., Furusaka, C. (1960) J. Biochem. (Tokyo)48, 831.
18. Hattori, T., Furusaka, C. (1961) J. Biochem. (Tokyo)50, 312.
19. Hulst, A. C., Tramper, J., Van't Riet, K., Westerbeek, J. M. M. (1985) Biotechnol. Bioeng. 27, 870.
20. Inada, Y., Takahashi, K., Yoshimoto, T., Ajima, A., Matsushima, A., Saito, Y. (1986) Trends in
21. Biotechnol. 4, 190.
22. Jack, T. R., Zajic, J. E. (1977) Adv. Biochem. Eng. 5, 125.
23. Jens6n, S., Westlake, D. W. S., Wolfe, S. (1984) Appl. Microbiol. Biotechnol. 20, 155.
24. Johansen, A., Flink, J. M. (1986) Biotechnol. Lett. 8, 121.
25. Kato, N., Yamagami, T., Shimao, M., Sahazawa, Ch. (1987) AppL MicrobioL Biotechnol. 25, 415.
26. Kaul, R., Adlerkreutz, P., Mattiasson, B. (1986) Biotechnol. Bioeng. 28, 1432.
27. Kay, G. (1968) Process Biochem. 3(8), 36.
28. Kennedy, J. F. (1982) Nature 299, 777.
29. Kennedy, J. F., Cabral, J. M. S. (1983) Solid Phase Biochemistry: Analytical Aspects, New York.
30. Kierstan, M., Bucke, C. (1977) Biotechnol. Bioeng. 19, 387.
31. Klein, J., Vorlop, K. D. (1985) Immobilization techniques -- cells. In: Comprehensive Biotechnology
32. 2, New York.
33. Klibanov, A. M., Samokhin, G. P., Martinek, K., Berezin, I. V. (1977) Biotechnol. Bioeng. 19, 1351.
34. Konecny, J. (1984) Swiss Biotechnol. 2(3), 15.
35. Kopp, B., Rehm, H. J. (1984) Appl. Microbiol. Biotechnol. 19, 141.
36. Ken, V., Ludvik, J., Kofrofiov, O., Kozovi, J., O, Ehfiek, Z. (1987) Appl. Microbiol. Biotechnol. 26,

37. 219.
38. Lee, C. K., Long, M. E. (1974) US Patent 3 821 086.
39. Leuscher, F. (1966) German Patent 1 227 855.
40. Manecke, G., Gunzel, G. (1966) Macromol. Chem. 91, 136.
41. Matulovic, U., Rasch, D., Wagner, F. (1986) Biotechnol. Lett. 8, 485.
42. Messing, R. A. (1975) Immobilized Enzymes for Industrial Reactors, New York.
43. Messing, R. A. (1980) Ann. Rep. Ferm. Proc. 4, 105.
44. Messing, R. A. (1985) Immobilization techniques -- enzymes. In: Comprehensive Biotechnology
45. 2, New York.
46. Mosbach, K., Mosbach, R. (1966) Acta Chem. Scand. 20, 2807.
47. Munton, T. J., Russel, A. D. (1973) Appl. Microbiol. 26, 508.
48. Clark, L. C. and Lyons, C. (1962) Electrode systerm for continuous monitoring of cardiovascular surgery Ann. NY Acad Sci 102,2945.
49. Guilbault, G G (1984) Analytical Uses of Immobilized Enzymes Marcel Dekker, New York.
50. Kaun, S. S. and Guilbault, G. G. (1987) Ion-selective electrodes and biosensors based on ISEs, m Biosensors. Fundamentals andApplicattons (Turner, A. P. F., Karube, I, and Wilson, G. S., (ed.), Oxford University Press, New York, pp. 135-152.
51. Kauffmann, J.-M. and Guilbault, G. G. (1991) Potentiometric enzyme electrodes, in Biosensor Principles and Apphcatrons (Coulet, L. J and Blum, P R (ed.), Marcel Dekker, New York, pp. 63-82.
52. Luong, J. H. T, Groom, C. A., and Male, K. B. (1991) The potential role of biosensors in the food and drink mdustrtes. Biosens. Bioelectron 6,547-554.
53. Brand, U., Brandes, L., Koch, V, Kullik, T., Remhardt, B., Ruther, F , Scheper,T., Schugerl, K., Wang, S., Wu, X., Ferrett, R., Prasad, S., and Wtlhelm, D (199 1) Monitoring and control of biotechnological production processes by Bio- FET-FIA-sensors Appl Microbrol Biotechnol 36, 167-1 72.

54. Ho, M H. (1988) Potenttomctric biosensor based on immobihzed enzyme membrane and fluoride detector Sens Act B 15,44545 1 8. Hmtsche, R., Dransfeld, I., Scheller, F., Pham, M. T., Hoffmann, W., Hueller, J.,
55. and Moritz, W (1990) Integrated differential enzyme sensors usmg hydrogen and fluoride ion sensitive multigate FETs. Biosens Bioelect 5,321-334
56. Guilbault, G G. and Luong, J. H T. (1989) Biosensors: current status and future possibilities. Selective Electrode Rev. 11, 3-16.
57. Beyer, M., Menzel, C., Quact, R., Scheper, T., Schiigerl, K., Treichel, W., Voit, H., Ullrich, M., and Ferret& R. (1994) Development and application of a new enzyme
58. sensor type based on the EIS-capacitance structure for bioprocess control. Bzosensor Blosensor Bioelect 9, 17-2 1.
59. Carr, P. W. and Bowers, L. D. (1980) Immobilized enzymes, in AnaZytzcal Chemistry' Fundamentals and Appkcations. Wiley, New York.
60. Gorton, L., Csoregl, E., Dominguez, E., Emneus, J., Jonsson-Pettersson, G , Marko-Varga, G., and Persson, B. (1991) Selective detection in flow analysis based on the combination of immobilized enzymes and chemically modified electrodes. Anal Chum Acta 250,203-248.
61. Cardow, M. F. and Turner, A. P. F. (1989) The realization of electron transfer from biological molecules to electrodes, in Baosensors: Fundamentals and Applecations (Turner, A. P. F., Karube, I., and Wilson, G., eds.), Oxford University Press, New York, pp 257-275.
62. Almelda, N. F. and Mulchandani, A. (1993) A mediated enzyme eletrode using tetrathiafulvalene and L-glutamate oxidase for the determination of L-glutamlc acid. Anal. Chum Acta 282,353-361.
63. Cass, A. E. G., Davis, G, Francis, G. D , Hill, H. A. O., Aston, W. J , Higgins, I. J , Plotkm, E. V., Scott, L D. L., and Turner, A. P. F. (1984) Ferrocene-mediated enzyme electrode for amperometric determination of glucose. Anal Chem. 56,667-67 1.
64. Kajiya, Y., Sugai, H., Iwakura, C., and Yoneyama, H. (1991) Glucose sensltlvity of polypyrole films containing immobilized glucose

oxidase and hydroquinonesulfonate ions Anal. Chem 63,49-54.

65. Schuhmann, W. (1993) Non-leaking amperometric biosensors based on high-molecular ferrocene derivatives. Blosens. Bioelect 8, 191-196.
66. Foulds, N. C. and Lowe, C. R. (1988) Immobihzation of glucose oxldase m ferrocene modified pyrrole polymers. Anal Chem. 60,2473-2476.
67. Heller, A. (1992) Electrical connection of enzyme redox centers to electrodes.J. Phys. Chem. 96,3579-3587.
68. Mulchandani,A.,Wang,C.-L.,andWeetall,H.H.(1955)Amperometrlc detection of peroxides with poly(anilinomethylferrocene) modified enzyme electrodes, Anal. Chem 67,94-100.
69. Wang, J., Wu, L. H., Lu, Z., Li, R., and Sanchez, J. (1990) Mixed ferroceneglucose oxidase-carbon-paste electrode for amperometric determination of glucose. Anal. Chum. Acta 228,251-257.
70. Pandey, P. C., Glazier, S., and Weetall, H. H. (1993) An amperometric flowinjection analysis biosensor for glucose based on graphite paste modified with tetracyanoquinodimethane. Anal. Blochem. 214,233-237
71. Mattiasson, B., Danielsson, B., Mandenius, C. F., and Winquist, F. (1981) Enzyme thermistors for process control. Ann. NY Acad. Scz. 369,295-305.
72. Mandenius, C. F., Danielsson, B., and Mattlasson, B. (1984) Evaluation of a dialysis probe for contmuous sampling in fermentors and in complex media. Anal.
73. Chcm. Acta 163, 135-141.

CHAPTER 9

IN SITU PRODUCT ISOLATION AND ASSOCIATED MICROBIAL PROCESSES

INTRODUCTION

The microbial products are generally removed from the microbial culture once the cultivation is terminated. Once the microbial products attain threshold concentration, they automatically inhibit the production of the microorganisms. The arrest of the microbial growth may occur when the product concentration is low that is less than even 10%. Some of the metabolites are antibiotics, organic acids such as lactic acid and salicycilc acid, solvents such as acetone and alcohols such as butanol and ethanol. *In situ* isolation of the products can reduce the waste water since the nutrients can be utilized at high concentrations. The operation costs will be reduced during this process. Some of the methods which are based on the microbe production and the composition of the media are filtration and crystallization, dialysis, ion exchangers, adsorption, extraction and evaporation.

EVAPORATION

In case of vacuum fermentation, the volatile products can be isolated from the culture media by creating vacuum inside the bioreactor. The products obtained are later evaporated at normal temperature. The product concentration is maintained at lower levels and hence it stops the product inhibition. There are three systems defined by Cysewski and

Wilke. These systems were aimed to produce ethanol from the substrates which are concentrated. If the glucose is supplied at constant rate into the bioreactor, ethanol production gets increased at a rapid rate and slowly the rate of production slows down after two days. The toxic and non-volatile products are reduced by isolating a fraction of culture from the bioreactor. Due to this technique, the production of ethanol increased by 38 %. If an additional recycling device of yeast is used, the production of ethanol is around 83 grams/liter/hour. Carbon dioxide should be compressed from the pressure at low level to the atmospheric pressure using sophisticated compressors. Limited amounts of the pure oxygen are supplied for the production of the yeast and the solubility of oxygen in the bioreactor is low at low pressures. In order to prevent these complications, vacuum fermentation is devised.

The bioreactor will run in the normal atmospheric pressure and the liquid circulated in the chamber with vacuum and the ethanol is volatized so that the inhibition will be kept at low rate. Due to separation of the bioreactor and the vacuum chamber, the carbon dioxide which is formed not necessarily is compressed. The cost of compression is drastically reduced and the air can be used in place of the pure oxygen. This system was used for the production of ethanol with the help of *Candida acidothermophilum* and *Zymomonas mobilis*. In case of ethanol reaching a concentration, which will start inhibiting culture, the culture fluid is allowed to pass between the bioreactor and the vacuum chamber until the concentration of the ethanol is dropped. After this step, the glucose is fed again. The volume of the bioreactor is restored after every ten cycles due to formation of the metabolites which are non-volatile. The vacuum fermentations alongside evaporation resulted in no difference from the traditional processes because the costs saved are compensated by the operation and investment costs. Limited applicability is seen in the process of vacuum fermentation. The product should be volatile and this volatility must be more than water and this property will restrict their application to acetaldehyde and ethanol production. The removal of culture from the bioreactor on continuously basis can be utilized to reduce the ill effects of byproducts which are non-volatile.

Extractive fermentation

In the process of fermentation, the extraction of the metabolites can be isolated from the fluid with a good water-insoluble solvent which is organic. The products which are dissolved in the solvent can be separated by the process of re-extraction into the buffer and distillation. The solvents and the culture fluid can be made in contact with each other either in the external extraction vessel and the bioreactor. This should be carried out with the help of the solvents which possess higher extractive capacity. These solvents should be selective to the products which are to be separated and they are non-toxic to the microorganisms. Finn was the first researcher who first used extractive fermentation. Good results were obtained when oleyl alcohol was used in the process of extractive fermentation. The ethanol extraction *in situ* was done by using a ethanol evaporation from the solvent and the solvent regenerated will be returned back to the system. The ethanol which is produced from the effluent is around 90%. The ethanol content present in the effluent is about 85 % and the rest of the volume is water. In this process, the system is stable and there will be negligible amount of toxic substances.

The oleyl alcohol can be used in extractive fermentation of butanol and acetone in a batch fed system. The fermentation process is same with the dibutyl phthalate which acts as a solvent. The important benefit of the compound is the stimulatory effect on the growth of the microorganisms. This process of extraction is used for the methane fermentation anaerobically in isolation of caproic acid, valeric acid and butyric acid. The extraction process can be done using kerosene. The acid concentration in the extraction solvent was 10 folds higher than the culture fluid. The propionic acids and acetone which is accumulated can limit the process of fermentation. Datta studied the effects of various solvents on the acidophilc and anaerobic bacteria. Toxicity was not found in the crude oil, amyl acetate and toluene which are concomitantly with trioctylphosphine oxide or a tertiary amine like Alamine 336. Smith and Playne tested 30 solvents on facultative anaerobic bacteria which produces organic acids. Out of the solvents tested, the nontoxic solvents are 12, and two solvents inhibited growth and remaining 15 were toxic to the microorganisms. The toxic effects on Lactobacillus delbreueckii can be seen in the sequence os alkane<cumene<Ketone<tertiary

amine<secondary amine<quaternary amine. The toxicity of the solvent basically depends on the microbial species used and the solvents which are to be suited should be selected by the process of screening.

Aqueous two-phase systems

In order to prevent problems which do arise when organic solvents are used in the extractive fermentations, the process of *in situ* product isolation from the culture fluid in presence of aqueous two-phase systems. In place of the organic solvents, the polymers were used till two phases are formed separately which consists of 95 % water and are biocompatible. The microorganisms exist on one phase and the low molecular weight substances are evenly distributed among both phases. The phase volumes can be altered so that the microbe containing phase is smaller than the other. Most of the product is devoid of the microorganisms. The best studied example is the toxin production using *Clostridium tetani* by the procee oṣ two aqueous phases. The solution which is used in this process comprises polyethylene glycol of 12 % and dextran of 2 % with the volume ratio among the two phases in the ratio of 15:1. The cells which rested in the tiny phase with dextran and the toxin are spread in both phases. The phase which comprises of polyethylene glycol (PEG) constitutes 15 folds more amount of toxin due to the favourable phase volume ratio. In the process of the batch extractive cultivation gave 100 times more production than the conventional process due to the toxic substances removal from the culture fluid.

The aqueous two-phase extractive fermentation can be used for production of the ethanol. The volume ratio of dextran which is on the top to the PEG which is at the bottom will be in the ratio of 9:1. The contents present in the bioreactor is stirred and the phase with dextran contains yeast which is in the form of droplets will be dispersed in the phase which contains PEG. Once the added glucose is consumed, the PEG phase which contains 90% ethanol is separated and is isolated by the process of distillation. The free PEG phase without ethanol is returned back to the bioreactor and again glucose is supplied afresh. Ten cycles of fermentation have yielded enhanced metabolic byproduct of glycerol and the toxic nonvolatile substances and their will be low rate of fermentation. Seperation of non-volatile metabolites by the process of dialysis and periodic addition of the yeast cells can restore the rate of the fermentation. This type of fermentation was tested on lab scale with other products also. In acetone and butanol production by *Clostridum acetobutylicum*, the extraction step reduced the product inhibition but the productivity was identical as the conventional fermentation which is not with the extraction step. This system comprised of 25% PEG and 6% dextran by weight. The system is the same for production of acetic acid by and the production enhanced by 50%.

This type of fermentation is also tested for biotransformation of several chemicals with the cells which are non growing. This method is termed as extractive bioconversion. This term was first given by Kaul and Mattiasson in the year 1986. In this process the top phase omprise of polypropylene glycol, methoxypropylene glycol and polyethylene glycol. The bottom phase consists of starch based polymers or dextran. In rare cases, buffer is made as bottom phase. The microorganisms which perform bioconversion are dispersed in the phase at the bottom and the products are deposited on the top phase and it is isolated. This system utilizes the differences in their hydrophobicity which is used for phase formation and the reactions occur in the higher concentrations of substrate. The extractive fermentation and the process of bioconversion in these phases are not well studied and hence the economic potential has not been assessed. The factors which are crucial are the polymer price, frequency of replacement, product isolation from the mixture of the polymers and byproduct accumulation whilst operation. An important adavantage of this process is that their biocompatibility with the polymers which allows the waste water volume. It is also useful in terms of broad applicability of products and byproduct accumulation and later isolation.

Extractive fermentation (pertraction)

In this type of fermentation, the microorganisms are accumulated in the liquid-liquid interface and it drastically reduces the rate of extraction. In order to avoid this complication, a novel method of extraction is devised and it is named as pertraction. In this method, the support membrane is liquid for extraction. The pertraction unit consists of a liquid support membrane and it is placed in the space between two spaces. This is reinforced with a plastic grid plate which is capable of causing turbulence. The culture grown can be allowed to pass from one side of the vessel and the sodium hydroxide solution removes the acids which are extracted acids and they pass to the remaining part of the vessel. Important parameter for a good pertraction extractive method is regulating the pH value of the fluid or culture fluid. The production capacity of this system is five times higher than the old process or the conventional process.

Adsorption

In situ desorption or adsorption basically on the charcoal was used successfully for isolating the products yielded from acetone-butanol fermentation from the culture fluid. This process is done by using the adsorbents which are porous and they do provide larger surface area such as polymeric resins and charcoal. The adsorbents can be added to the bioreactor or can be installed in a vessel which is located outside and the culture fluid is allowed to pass through it. This process is used for producing ethanol. The charcoal is deposited into the bioreactor. The production is not improved though there is decrement in the free ethanol concentration. The adsorption process affects the yeast cell viability. Good results were obtained when the non-growing and stationary yeast cells were isolated from the culture fluid and so there are excluded from its contact with the charcoal surface. Once ethanol is separated, the culture medium and yeast cells are returned back into the bioreactor. The process of fermentation is allowed to continue after the addition of glucose. The productivity of ethanol by this process is 25 g/l/h.

The *in situ* isolation of ethanol by the process of fermentation has been tried with various hydrophobic adsorbents by Glaser, Einicke and Schollner. Liquid phase experiments showed that the pentasil zeolites do have enhanced selectivity for ethanol even in least concentration range. If the fermentation broth gets contacted with zeolite NaZSM-5, then there would be no inhibition of the product. The production rate of ethanol depends on the adsorbent to glucose ratio. The *in situ* isolation of isopropanol and butanol from *Clostridium beyerinckii* can be done by adding polymeric resins and charcoal to the culture. After completion of the cultivation, the adsorbents are isolated from a metal filter. The products were desorbed at 150 ^{0}C and these adsorbents can be used for adsorption repeatedly. Good results are obtained when cycloheximide is isolated from the culture fluid by the process of *in situ* adsorption. The antibiotic synthesis from Streptomyces griseus is feed-back regulated by the product. The culture fluid is allowed to pass throughout the process of fermentation via polymeric resin. This causes a decrement in the concentration of cycloheximide in the culture. This process yielded overall production of 1600 μg/ml. If ultramembrane is coated with resin, the concentration is enhanced to 1800 micrograms per ml. It shows that the resins can inhibit the production. The toxic nature of the resins on microorganisms is one of the serious problems in case of *in situ* isolation of the product. Some of the factors which make this process complicated are resin abrasion, intermediates, medium components, unspecific absorption of cells and adsorbents used for isolating products from the culture. Factor such as toxicity can be reduced by using immobilized cells inside the bioreactor or by isolating them from the fluid well before the point of contact with the asorbent. These modifications will enhance the complexity and the costs will rise up.

Ion exchangers

The ion exchangers are packed into a column and the culture fluid flows into it or they can be added directly into the culture fluid. In the process of salicylic acid production from naphthalene by *Pseudomonas aeruginosa*, 10 g/l salicylic acid do provide toxic effect on the cells and thereby product inhibition occurs and also degradation of salicylic acid. If anion exchanger resin is added to the culture fluid, there is an increase of salicylic acid concentration which amounts to 30 g/l. When the resin is subjected to encapsulation by using a cellophane membrane and then allowed to add into the bioreactor and thereby avoiding contact with the resin, the inhibitory effect was reduced and the production achieved is around 55 g/l. The salicylic acid produced by *Corynebacterium renale* using this ion exchange column gave 14.6 g/l of salicylic acid. It is double the yield than the conventional process. The ion exchange resins were also utilized for isolation of antibiotics such as neomycin and novobiocin from the culture fluid. The ions which are released from the ion exchanger can inhibit the growth and the ionic compounds can be adsorbed on the surface of the resin and can even reduce or dimish the adsorption of the product. The metabolites which are non-ionic can also be accumulated in the culture broth. The ion exchanger resins are costly and the mass transfer is considered as slow. The capital costs are higher and the devices are large. These factors can decrease productivity and so this process is considered as not economic.

Dialysis

In this type of fermentation, the semipermeable membrane demarcates the medium reservoir from culture space. The nutrient which is from the reservoir gets diffused into the culture liquid and the metabolites get diffused into the reservoir from the culture. Always lower concentration of product is maintained in the space where culture is present so as to avoid the inhibitory effects. This fermentation was well studied in lactic acid production process. Friedman and Gaden proposed that lactic acid produced via this method produced 8 g/l and it is better than the conventional process which yielded 5 g/l. Gerhardt and Stieber devised two methods for the fermentation of lactic acid using dialysis fermentation. In once system, the substrate is allowed to get into the bioreactor continuous circuit which is supplied with water at its pure state. High conversion of substrates and speedy growth of cells is possible with this method. The effluent contains product and is devoid of cells.

The continuous culture with dialysis can be represented by a system. This system is with the substrate which is fed into a dialyzer circuit and again into the bioreactor circuit through dialysis. If the bioreactor circuit does not have an outlet that is batch configuration and it does contain cells. The products are removed continuously via dialysis. One of the advantages is that the substrate is convereted into the product without any fresh addition of substrate. The outlet can be installed both in the bioreactor and the dialyzer or in sometimes it can be installed in dialyzer circuit only.

Solid phase naphthalene is used as substrate and the salicylic acid is allowed to be diffused into the reservoir in case of semicontinuous dialysis system which is utilized in salicylic acid production from *Pseudomonas fluorescens*. If the salicylic acid concentrations hit the inhibitory level, then fresh medium is added from the reservoir. Though the production rate of salicylic acid is 2.6 times greater than the conventional methods, the concentrations are lower. Cycloheximide extraction from Streptomyces griseus uses cellophane dialysis tube. This molecule is extracted from water with the help of methylchloride and the remaining water is returned back to the dialysis circuit. This technique has resulted in double yields of antibiotic concentration. The primary advantage of the dialysis fermentation process is that there cannot be inhibition due to the end product and higher amounts are cells are being holded up and therefore high cell concentrations. Some of the limitations are higher costs of product isolation, clogging of membranes and deposition of metabolites which are non-diffusible and thus inhibits the growth and metabolism.

Filtration

It is a good method for isolating and extracting the inhibitory metabolites from the system. Several types of hollow fibre bioreactors were constructed for the purpose stated above. It is a cyclinder and it is divided into two spaces with the help of asymmetric membrane and hollow fibres. The membranes possess anisotropic structure and it is composed of ultramicroporous inner layers which are enclosed in a matrix which is macroporous and made up of polymer. The inner dense layer can act as semipermeable membrane and the macroporous matrix which offers 60 to 90 % void space gives higher rate of hydraulic permeability. The medium is allowed to pass into the inner bioreactor space and the medium diffuses through the ultramicroporous membrane. The metabolic products which are excreted diffused into another direction and are successfully isolated. The fermentation with the help of the hollow fibre filtration is being tested with many processes. The synthesis of the beta lactamase by the *E. coli* is one of the examples for this process where there will be production of ethanol, superoxide dismutase and lactic acid. By this process, the cell mass is about 19 g/l and four times higher production of superoxide dismutase than conventional methods. Though it seems that the hollow fibre macrofiltration is hard to sterilize, the flow rate of the produced filtrate via hollow fibres are very low. In many practical purposes, the inhibitory metabolites are separated using ceramic materials which offer higher mechanical stability and are resistant to higher temperatures and chemicals. This type of properties is essentially desirable in the fermentations which do require washing and steam sterilization. There are some problems but the hollow fibre modules were used for much fermentation especially in lactic acid fermentation. The efficiency of this ystem was tried and tested with producers of lactic acid such as *Lactobacillus casei* and *Streptococcus cremoris*. The cell concentrations achieved are 81 g/l and 49 g/l respectively by their dry weight.

Crystallization

The isolation of product with the method of crystallization has only been used in biotransformation of L-aspartate to L-alanine with the help of immobilized *Pseudomonas dacunhae*. This system comprises of a bioreactor which contains immobilized bacteria and the effluent is obtained from the separator and crystallizer. Post product separation, the filtrate which contained substrate will be passed into the reservoir which is substrate feed. The substrate is fed afresh again and then returned back to the bioreactor in order to start bioconversion cycle. This process has yielded good improvement in the quality of the product and the costs of production were reduced to 25%. These systems can be used for various product operations.

REFERENCES

1. Abbott, B. J., Gerhardt, P. (1970) Bioteehnol. Bioeng. 12, 577.
2. Cohen, A., van Gemert, J., Yoetemeyer, R. J., Breure, A. M. (1984) Process Biochem. 19, 228.
3. Crabbe, P. G., Tse, C. W., Munro, P. A. (1986) Biotechnol. Bioeng. 28, 939.
4. Cysewski, C. R., Wilke, C. R. (1977) Biotechnol. Bioeng. 19, 1125.
5. Datta, R. (1981) Biotechnol. Bioeng. 23, 61.
6. Daugulis, A. J., Swaine, D. E., Kollerup, F., Groom, C. A. (1987) Biotechnol. Lett. 9, 425.
7. Denkewalter, R. G., Gillin, J. (1959) German Patent 1 062 891.
8. Einicke, W. D., Glaser, B., Sch611ner, R. (1991) Acta Biotechnol. 11, 353.
9. Finn, R. K. (1966) J. Ferment. Technol. 44, 305.
10. Friedman, M. R., Gaden, E. L., Jr. (1970) Biotechnol. Bioeng. 12, 961.
11. Ghose, T. K., Roychoudhury, P. K., Ghose, P. (1984) Biotechnol. Bioeng. 26, 377.
12. Groot, W. J., Luyben, K. C. A. M. (1986) Appl. Mierobiol. Biotechnol. 25, 29.
13. Holst, O., Hanssson, L., Berg, A. C., Mattiasson, B. (1985) Appl.

Microbiol. Bioteehnol. 23, 10.
14. Inloes, D. S., Smith, W. J., Taylor, D. P., Cohen, N. S. (1983) Biotechnol. Bioeng. 25, 2653.
15. Ishii, S., Taya, M., Kobayashi, T. (1985) J. Chem. Eng. Japan 18, 125.
16. Kaul, R., Mattiasson, B. (1986) Appl. Microbiol. Biotechnol. 24, 259.
17. Kitai, A., Tone, H., Ishikura, T., Oyaki, A. (1968) J. Ferment. Technol. 46, 442.
18. Kominek, L. A., (1975) Antimicrob. Agents Chemother. 7, 856.
19. Kuhn, I. (1980) Biotechnol. Bioeng. 25, 2393.
20. Lee, Y. H., Chang, H. N. (1989) Biotechnol. Lett. 11, 23.
21. Lee, S. S., Wang, H. Y. (1982) Biotechnol. Bioeng. Symp. 12, 221.
22. Lencki, R. W., Robison, C. W., Moo-Young, M. (1983) Biotechnol. Bioeng. Syrup. 13, 617.
23. Levy, P. F., Sanderson, J. E., Wise, D. L. (1981) Biotechnol. Bioeng. Symp. 11, 239.
24. Maiorella, B., Blanch, H. W., Wilke, C. R. (1983) Biotechnol. Bioeng. 25, 103.
25. Mattiasson, B., Siominen, M., Andersson, E., Haggstrom, L., Albertson, P.-A., Hahn-Hagerdal, B.
26. (1982) In: Enzyme Engineering.
27. Mattiasson, B., Hahn-H.gerdal, B. (1983) In: Immobilized Cells and Organelles 1, 121.
28. Mehaia, M. A, Cheryan, M. (1986) Enzyme Microbiol. Technol. 8, 289.
29. Minier, M., Goma, G. (1982) Biotechnol. Bioeng. 24, 1565.
30. Nishizawa, Y., Mitani, Y., Tamaki, M., Nagai, S. (1983) J. Ferment. Technol. 61, 599.
31. Nishizawa, Y., Mitani, Y., Fukunishi, K., Nagai, S. (1984) J. Ferment. Technol. 52, 41.
32. Ohleyer, E., Blanch, H. W., Wilke, C. R. (1985) Appl. Biochem. Biotechnol. 11, 317.

33. Playne, M. J., Smith, B. R. (1983) Biotechnol. Bioeng. 7, 1251.
34. Puziss, M., Hed6n, C.-G. (1965) Biotechnol. Bioeng. 7, 355.
35. Schlosser, S., Kossazcky, E. (1986) J. Rad. Nucl. Chem. 101, 115.
36. Stieber, R. W., Gerhardt, P. (1981a) Biotechnol. Bioeng. 23, 523.
37. Stieber, R. W., Gerhardt, P. (1981b) Biotechnol. Bioeng. 23, 535.
38. Takamatsu, S., Ryu, D. D. Y. (1988) Enzyme Microb. Technol. 10, 593.
39. Tangu, K. S., Ghose, T. K. (1981) Process Biochem. 16(2), 24.
40. Tanigushi, M., Kotani, M., Kobayashi, T. (1987) J. Ferment. Technol. 65, 179.
41. Taya, M., Ishii, S. Kobayashi, T. (1985) J. Ferment. Technol. 63, 181.
42. Wang, H. Y., Kominek, I. A., Jost, J. L. (1980) In: Advances in Biotechnology 1,601.
43. Wang, H. Y., Robinson, F. M., Lee, S. S. (1981) Bioteehnol. Bioeng. Symp. 11, 555.
44. Wang, H. Y. (1983) Ann. N. Y. Acad. Sci. 413, 313.
45. Wayman, M., Parekh, R., (1987) J. Ferment. Technol. 65, 295.
46. Weizmann, C., Bergman, E., Sulzbacher, M., Pariser, E.R. (1948) J. Soc. Chem. Ind. 67, 225.

INDEX